CREATIN

INTANGIBLE ENTERPRISE

"In *Creating the Intangible Enterprise*, KP emphasizes that paradigm-shifting technologies like AI won't upend the world as we know it. On the contrary, they will bring us back to the fundamentals—enhancing success through greater creativity, adaptability, resilience, and leadership. It's ultimately a hopeful vision of a future that's more efficient and more human."

— Aamir Paul, President for North America and Member of the Executive Committee, Schneider Electric

"In his own inimitable way, with tremendous insight and a dash of humor, Reddy lays out a clear vision for an AI-infused future, and a road map for design-firm CEOs to thrive in it."

— Mick Morrissey, Managing Principal, Morrissey Goodale

"Intelligent, mindful, and fun. Rigorous yet balanced with levity through humor and personal anecdotes. A great job of disarming the assaults on AI and positioning it as the valuable asset it really is—an enhancement, not replacement, for human creativity, judgment, and decision-making. A lot of leaders, strategists, and anyone uncertain about the role of AI should rush to grab a copy!"

**— Barry Johnson, Chief Executive Advisor,
BEAJ GLOBAL, Inc.**

"KP gives the reader a highly personalized, deeply engaging, and brilliant analysis on the role AI will play—and won't play—and how professionals need to best prepare themselves for the coming AI revolution."

— Michael Beckerman, CEO of CREtech

"*Creating the Intangible Enterprise* simplifies the way we should look at the rise of AI use cases and provides insightful, practical recommendations to be ready and thrive as the organization of the future."

**— Ibrahim S. Odeh, Ph. D, MBA, Department of Civil
Engineering and Engineering Mechanics,
Columbia University**

THE CRITICAL SKILLS REQUIRED
TO THRIVE IN AN AI-DRIVEN WORLD

CREATING THE
INTANGIBLE
ENTERPRISE

KP REDDY

First printing 2024

Book cover by Burtch Hunter
Book interior by Najdan Mancic

ISBN 979-8-9902814-1-7 Paperback
ISBN 979-8-9902814-2-4 Hardback
ISBN 979-8-9902814-0-0 Ebook

Published by Ripples Media
www.ripples.media

DEDICATION

My books are projects of passion that become all consuming. Nights, weekends and some vacations. While all along the way being filled with doubt. My family not only tolerates my distractions during this process, they support me in ways that are extraordinary. Thank you Rachel, D1, J2 and R3. You all support me in ways that you may not even realize.

TABLE OF CONTENTS

PREFACE

Throughout this book, I spend a significant amount of time explaining how artificial intelligence will absolutely change the world, but perhaps not in the ways that we think it will. In the following pages, I will unpack how other technological advances were seemingly poised to wipe out entire industries—only they didn't.

For instance, the creation of spreadsheets was supposed to wipe out the entire accounting field. Alas, the accountants are doing quite well. As I will explain, adopting spreadsheets necessitated a need for accounting departments to consult on implementing this technology. The solution wasn't choosing between accountants and spreadsheets, but rather integrating accountants *and* spreadsheets. There was a human need to guide and oversee adoption and implementation.

Not to give away the ending, but the core tenet of this book is that we should view artificial intelligence through the lens of humans *and* algorithms, not humans *or* algorithms. The opportunity lies in humans possessing the knowledge and creativity to best utilize AI so that companies can efficiently solve problems and increase revenue.

I could have had AI write this entire book. AI software like ChatGPT can very capably define terms like deep learning, computer vision, and expert systems, as well as cite specific examples of how AI is utilized in different industries like agriculture, medicine, law, etc. However, AI can't contextualize these examples from my perspective and give relevant takeaways for how you should look at this powerful technology. Plus, you the reader would've also missed out on a wealth of sarcasm (stay tuned for a handful of jokes at the expense of civil engineers), as well as stories from my unique upbringing (like why I wasn't allowed to use erasers while doing math problems).

So, how did I utilize AI to help me write this book? Mainly, I asked for specific examples and definitions—like you'd use a Google search. In sections where I directly quoted or paraphrased responses from ChatGPT, those sections are cited as endnotes, and then the prompts are listed in the appendix.

I believe that in the future, AI will become a useful organizational and background tool for many writers. Still, the success of any book is entirely dependent on the writer's voice, ideas, and analysis.

In the tech world, we like to use the phrase, "eat our own dog food." By this, we mean that we like to use the ideas we sell. As you digest this book, view it through the same practice—it is an exercise to use "skills of the future" to write it.

Along with working extensively with my team of (actual human) editors, the other very human area is explaining the concepts to a broad audience. The flow and context of these ideas were very personal. I also spent months talking to hundreds of executives to determine if there was a market need and whether my hypothesis of the skills for the future was relevant.

This deeply personal project was the result of that—an exercise in using human analysis to test and evaluate the functions of AI. My mission was to focus on the opportunities of AI versus the fears of AI. I felt that it was best accomplished by giving some background and history of technologies that promised massive disruption, showing how people adapted to this technology, and providing some focused strategies and skills to thrive in an AI-driven world.

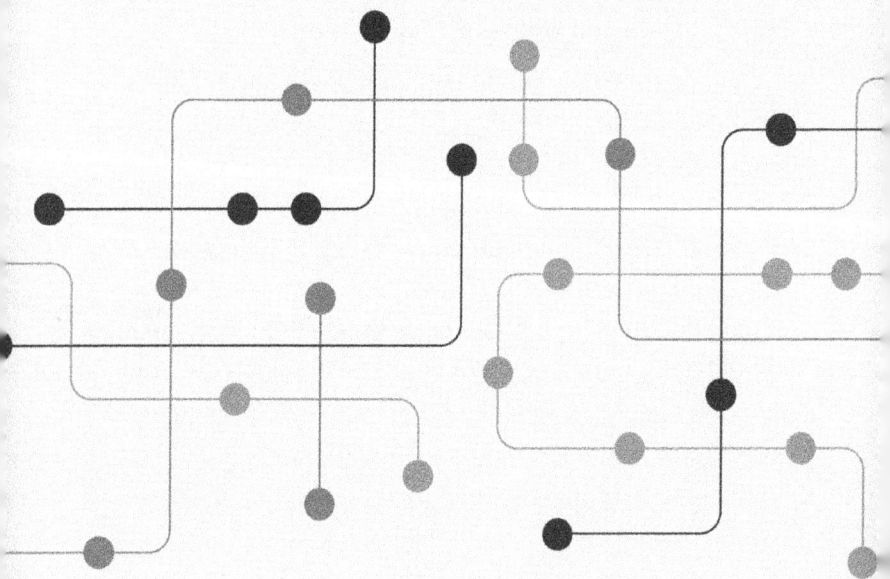

PART ONE

HOW DID WE GET HERE?

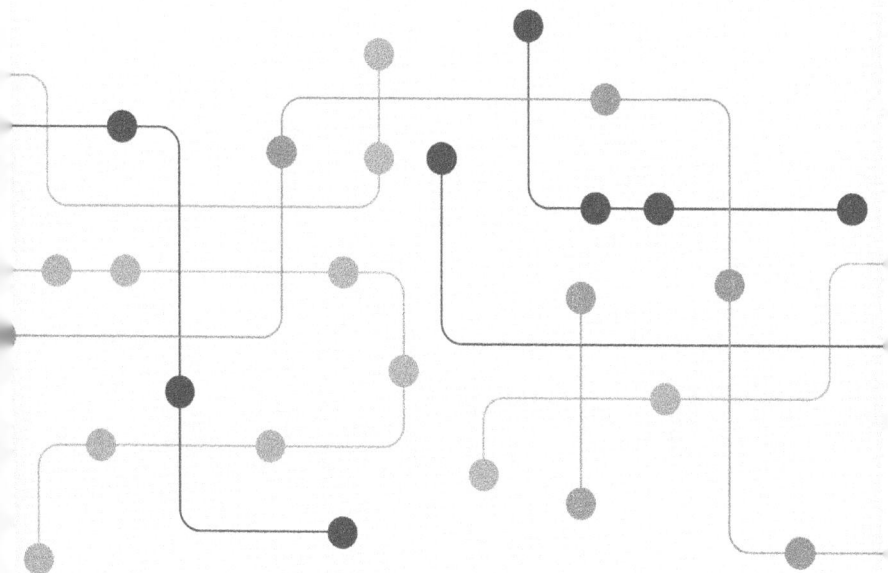

CHAPTER 1

EARLY ADOPTERS ARE REWARDED

head fake can happen in many things, but for me, it's most commonly a basketball reference. I'm 5'8" with the right shoes, so my basketball experience with head fakes is quite limited. Let's just say I have heard about them on TV. In technology, there is the constant head fake of "this will change everything."

I have always been an early adopter. I had a pager in the early days and I wasn't even a drug dealer. I had a two-way pager, the RIM Pager (predecessor of the BlackBerry), and then the BlackBerry, iPhone, etc. Each new device came with the promise of "changing everything."

During the early days of the Internet, I started figuring out how to build and tinker with websites. It wasn't hard per se, but as this was 1991, it wasn't exactly easy. When I graduated from college and got my first corporate job at an engineering company, I assumed all my co-workers had a computer. It turned out that very few of the engineers had computers, as they were reserved for the accounting department and the steno pool. What was the steno pool, you ask? It was the department of wonderful people who transcribed our handwritten or recorded notes into WordPerfect, printed them out, and left them in our physical mailboxes.

While at this engineering firm, I started a side business building websites for small companies. It wasn't hard to see that I could make more money doing that than I was earning as a civil engineer.

When I eventually quit, my immediate boss asked me what had taken so long, figuring I would have left much earlier. Many of the other engineers thought I was making the biggest mistake of my life. They believed that the world would always need civil engineers, but the "Internet" was just a fad that was going nowhere.

After I resigned, a colleague followed suit the next week. He and I believed in what the Internet could be and we shared a vision that the construction industry could be managed by utilizing this new tool. As we bootstrapped our new venture, we quickly discovered that the construction industry did not in fact share this vision. They deemed the Internet a toy that very serious engineers had no use for. Fortunately, we found other customers in the telecom industry who thought differently and served as our initial customer base.

Keep in mind that in 1997 the Internet involved dial-up modems, limited connectivity, and was far from a frictionless user experience. Most people were just working through a steady stream of free AOL CD mailer credits and thought the "Internet" was the same as AOL. As we built a reputation, we fielded phone calls from people asking us how they could get their own "dot-com."

Eventually, our company, Cereus Technology Partners, grew and went public in 1998. In many ways we got lucky, but our initial instincts were right. Even in the Internet's infancy, we knew it would be a game changer. The dot-com boom and bust left many believing it was still a short-lived fad. Like the civil engineers thought, who in their right mind would buy goods off the Internet? Another take that didn't age well.

The point here is to remind us that new technology is often met with skepticism. Executives and decision-makers may resist trying something new until they are forced to. In these cases, the early adopters and believers are rewarded. It is not difficult to draw parallels between the early stages of the Internet and the current wildfire spread of artificial intelligence (AI).

A BRIEF HISTORY OF TECHNICAL PROGRESSION

It seems preposterous now, but there was a time when emails addressed to a CEO were routed to their assistant who physically printed them and put them on the CEO's desk. In most cases, the CEOs were the biggest resistors to change. Then came the BlackBerry that changed everything. CEOs could now be at an "offsite" meeting (the golf course) and still be able to respond to emails. CEOs then flew into IT departments demanding a BlackBerry.

BlackBerry adoption took off like a rocket. Not since the BlackBerry have we seen such engagement of a new technology by the C-Suite. CEOs now have ChatGPT on their phone and are not only using it but are imagining its potential future utilization. They are now marching into IT departments and demanding an AI strategy. This top-down enthusiasm matched with curiosity does indeed "change everything."

We can think of technology progression in three stages: desktop programming (1980s), visual programming (2000s), and AI programming (2010s).

Desktop Programming

The evolution of desktop software programming has made the development process easier. In this section, ChatGPT was used to source some of these examples. Here are some factors that have contributed to this.

► **High-Level Languages and Abstraction—** "Introducing high-level programming languages, such as C, C++, Java, and C#, gave developers more abstract and expressive tools to write code. These languages abstracted away low-level details, such as memory management and hardware-specific operations, making programming more accessible to a broader range of developers. This abstraction allowed developers to focus more on solving problems and implementing functionality rather than getting bogged down in low-level implementation details.

▶ **Integrated Development Environments (IDEs)—** "IDEs have significantly improved the productivity of desktop software developers. IDEs provide a comprehensive set of tools, including code editors, compilers, debuggers, and code analyzers, that are all integrated into a single development environment. This integration streamlines the development process, automates repetitive tasks, provides real-time feedback and suggestions, and simplifies debugging and testing. IDEs, like Visual Studio, Eclipse, and JetBrains, have become indispensable tools for desktop software development.

▶ **Frameworks and Libraries—** "The availability of robust frameworks and libraries has accelerated the development process. Frameworks provide pre-built components, modules, and abstractions that developers can leverage to build applications more efficiently. These frameworks handle common functionalities such as user interface design, networking, database interaction, and security, allowing developers to focus on the core application logic rather than reinventing the wheel. Popular desktop software frameworks include .NET, Qt, Electron, and WPF (Windows Presentation Foundation).

► **Code Reuse and Modularity—** "Object-oriented programming (OOP) principles and modular design approaches have made it easier to write reusable and maintainable code. OOP allows developers to encapsulate code into reusable objects and classes, promoting reusability and reducing code duplication. Modular design further enhances code organization by breaking down large applications into smaller, manageable modules or components. This modular approach simplifies development, testing, and maintenance, as each module can be developed, tested, and updated independently.

► **Documentation and Community Support—** "The availability of extensive documentation, tutorials, and online communities has made it easier for developers to learn and troubleshoot programming challenges. Developers can access online resources, forums, and communities to find solutions to common problems, learn best practices, and stay updated with the latest trends and techniques. This wealth of knowledge and support helps developers overcome obstacles and accelerate their learning and development process.

► **Automated Testing and Continuous Integration**—"The adoption of automated testing practices and continuous integration (CI) has improved the quality and reliability of desktop software. Automated testing frameworks and tools allow developers to write test cases that can be automatically executed, reducing the need for manual testing and catching bugs early in the development cycle. Continuous integration allows developers to build, test, and deploy software automatically as changes are made, ensuring that the application remains working and facilitating collaboration among team members.

► **Cloud-Based Services and Infrastructure**—"The availability of cloud-based services and infrastructure has simplified various aspects of desktop software development. Cloud platforms offer scalable and reliable storage, hosting, and computing resources, eliminating the need for developers to manage and maintain their own infrastructure. Cloud services also provide ready-to-use functionality, such as authentication, file storage, and data analytics, which developers can integrate into their applications with minimal effort."[1]

Visual Programming

No-code, also known as "visual programming" or "low-code," refers to a software development approach that allows users to create applications without writing traditional programming code. Instead of manually writing code, users can use intuitive graphical interfaces and pre-built components to design and configure their applications visually.

In a no-code environment, users typically work with drag-and-drop interfaces, forms, and visual elements to define the application's structure, logic, and functionality. They can specify workflows, data models, user interfaces, and integrations by visually connecting different components or blocks. These components represent actions, conditions, data sources, and user interface elements. In this case, we can again look to ChatGPT for examples.

No-code platforms often provide a range of pre-built templates, widgets, and connectors to common services, databases, and APIs. This allows users to rapidly assemble applications by selecting and configuring these pre-existing components rather than writing code from scratch. The platforms usually abstract away the underlying complexities of the code, making it accessible to users with little-to-no programming experience.

No-code development has gained popularity due to its potential to democratize software development, enabling non-technical users, such as business analysts or citizen developers, to create functional applications. It allows them to solve specific business problems or automate workflows without relying on dedicated software development teams.

The benefits of no-code development, which we're also calling "visual programming," include:

▶ **Accessibility**— "No-code platforms provide a visual and intuitive interface that lowers the entry barrier for creating applications, making software development accessible to a broader audience. Users can create applications without coding knowledge or experience, empowering them to bring their ideas to life.

▶ **Speed and Efficiency**— "With pre-built components and visual interfaces, no-code development allows for rapid application development. Users can quickly assemble and configure components, reducing the time and effort required to build applications compared to traditional coding approaches.

▶ **Flexibility and Iteration**— "No-code platforms often provide flexibility for users to iterate and refine their applications rapidly. Changes can be made visually, allowing for quick experimentation and adjustments to the application's logic, user interface, or data model.

▶ **Integration Capabilities**— "Many no-code platforms offer built-in connectors and integrations with various services, databases, and APIs. This simplifies the process of connecting and interacting with external systems, enabling users to create applications that interact with other software or data sources seamlessly.

▶ **Empowering Citizen Developers**— "No-code development empowers non-technical users, known as citizen developers, to take an active role in creating applications. It enables them to address specific business needs independently, reducing dependency on traditional IT departments and fostering innovation at various levels within organizations."[2]

However, it's important to note that no-code development does have some limitations. Complex or highly customized applications may still require traditional coding approaches for greater flexibility and control. Additionally, the visual nature of no-code platforms might impose certain limitations on advanced programming concepts or specific use cases that require low-level optimizations.

AI Programming

In a similar evolution to what we've seen with visual or "no-code" programming, AI plays a significant role in accelerating software development. Here are a few examples, according to Chat GPT.

► **Code Generation—** "AI can generate code based on high-level specifications, reducing the manual effort required for implementation. For instance, tools like Chat GPT-3 can generate code snippets or even complete functions based on natural language descriptions.

► **Automated Testing—** "AI can help automate testing processes by generating test cases, identifying potential bugs, and even suggesting fixes. This reduces the time and effort spent on manual testing and improves overall software quality.

Bug detection and resolution: AI techniques, such as static code analysis and machine learning, can detect and fix software code bugs. These techniques can analyze large codebases, identify patterns, and suggest bug fixes or optimizations.

► **DevOps and Continuous Integration/ Continuous Deployment (CI/CD)—** "AI can enhance DevOps practices by automating various stages of the software development lifecycle. AI-powered systems can automatically build, test, and deploy software, making the development process more efficient and error-free.

► **Natural Language Processing (NLP) for Requirements Engineering—** "AI-based NLP techniques can help extract and understand requirements from natural language documents. This helps software developers better comprehend user needs and translate them into software features, reducing ambiguity and improving software design.

► **Predictive Analytics—** "AI can analyze vast amounts of data generated during the software development process, such as code repositories, bug databases, and user feedback. By identifying patterns and trends, AI can provide insights to optimize development practices, estimate project timelines, and improve resource allocation.

► **Intelligent Code Editor—** "AI-powered code editors can provide intelligent code completion, error detection, and code refactoring suggestions while developers write code. These tools enhance developer productivity by speeding up coding tasks and reducing syntax errors."[3]

It's important to note that while AI can automate certain aspects of software development, human expertise and judgment remain crucial. AI is most effective when it augments human developers, helping them work more efficiently and effectively.

CHANGING EVERYTHING. AGAIN.

Our biggest takeaway from the technological advances of the last fifty years is that we've been here before. AI is new in the grand scheme of things, but paradigm-shifting advances are most certainly not. In each of these evolutions, the technological shifts reduced the barriers to entry for professionals. The work that had once been done only by specialists could now be done by the masses. With the advent of visual programming, entire websites and applications can now be built by an individual who has never written a line of code—much less someone who majored in computer science.

Yet, with all of this in mind, there is still plenty of work for developers. Sure, they're not needed to build simple websites or landing pages, but they are still very much needed for more complex and technical projects.

AI is the next step in technology's ability to reason, process, learn, and simulate. With ChatGPT on your phone, you can now ask AI to draw "a gorilla on a seesaw," and the technology will produce a rendering that is on par with what a professional creative could design. You can ask it to produce a paper on a given topic (let's use the Vietnam War here). In seconds, you will have a paper outlining the complexities of Cold War tensions, North-South divide, and colonial history that would've taken an undergraduate history major multiple hours to research and write.

AI is a tool that will fundamentally change how companies work and how executives make decisions because they now have access to technology that can accomplish tasks previously only able to be done by humans. While the technology's applications can take any number of forms, there is one salient takeaway: operational efficiency will increase as companies utilize technology to accomplish tasks and deliverables in minutes, which would take employees far longer.

Almost every organization reaches a point where metrics drive many, if not all decisions. These metrics take names like profit margin, EBITDA (Earnings Before Interest, Taxes, Depreciation, and Amortization), KPI (Key Performance Indicators),

OKR (Objectives and Key Results), and the like. The financial metrics especially are the gauge used to measure enterprise value. Management, more specifically, doesn't manage the business but manages the metrics. This management of metrics involves a lot of report and presentation creation. With the advent of easily adoptable technology and new AI applications being developed daily, these metrics have the potential to be normalized. The best practices for operations powered by AI will level the playing field and enterprise values will all move to the mean.

THE *MONEYBALL* EFFECT

Let's look at a specific example. Michael Lewis's bestselling book *Moneyball: The Art of Winning an Unfair Game,* written in 2004, follows the progress of Oakland Athletics general manager Billy Beane. Beane was the longtime front office executive of the A's, a club that routinely had one of the lowest payrolls in Major League Baseball. Nonetheless, the A's were one of MLB's most consistent winners in the early 2000s, as Beane used analytics to find undervalued players, emphasizing statistics like on-base percentage that were mostly ignored by other front offices at the time. Other teams relied on the experience of scouts to develop winning teams, while Beane prioritized data and analytics.[4]

On a shoestring budget from 2000 to 2006, the A's enjoyed seven straight winning seasons, a winning percentage of .586, and made the playoffs five times. It was breakthrough thinking in an industry where team-building and player evaluation practices had remained largely unchanged for the last fifty years and was a clear differentiator for the franchise. That is, until everyone had the "Moneyball" secret formula. Lewis' book was published in 2004 and shortly after that, every ballclub utilized sabermetrics. Teams had shifted their strategy, at least in some form, from picking talent through scouts' recommendations to picking talent based on data and analytics. Beane's secret recipe was no longer an advantage once everyone had the formula. On cue, from 2007 to 2011, the A's dropped to a winning percentage of .470, missed the playoffs all five years, and did not have a single winning season.

ABUNDANT AI

This normalizing effect is what advanced technology and AI are bringing to bear, not just to innovators but to the mainstream. AI is not new—it is just now accessible to the masses, both from an affordability and a capability perspective.

In 1950, Alan Turing developed a thing called the imitation test. Basically, he wanted to figure out if a computer could be so smart that it could fool a human into thinking it's also a human.

He came up with this experiment where a person would have conversations with both a computer and another human without knowing which is which. If the person couldn't tell the computer from the human, the computer was considered "super smart" and passed the imitation test.

Turing believed that if a computer could convincingly imitate a human, then it must have some kind of intelligence. He thought this was a good way to determine if a machine could think like a human.

The imitation test became famous, giving way to what's now known as the Turing Test. It's still used today to measure the "smartness" of AI systems. People try to create AI that can chat with humans in a way that feels natural and human-like.

Before the year 2000, several significant milestones were achieved in the field of AI. Here are a few notable ones, according to ChatGPT.

► **The Dartmouth Workshop**— "In 1956, the Dartmouth Workshop, organized by John McCarthy, Marvin Minsky, Nathaniel Rochester, and Claude Shannon, is considered the birth of AI as a field. It brought together researchers who coined the term 'artificial intelligence' and laid the foundation for future developments.

► **Logic Theorist—** "In 1956, Allen Newell and Herbert A. Simon developed the Logic Theorist, the first AI program capable of proving mathematical theorems. It demonstrated that computers could perform human-like reasoning tasks.

► **General Problem Solver—** "In 1957, Allen Newell and Herbert A. Simon created the General Problem Solver (GPS), an AI program designed to solve a wide range of problems by searching through possible solutions. It was a significant step towards developing general-purpose problem-solving systems.

► **ELIZA—** "Developed in the mid-1960s by Joseph Weizenbaum, ELIZA was an early example of a natural language processing program. It simulated a conversation with a psychotherapist and showcased the potential of AI in understanding and generating human-like language.

► **Shakey the Robot—** "In the late 1960s, the Stanford Research Institute developed Shakey, one of the first mobile robots capable of perceiving its environment and performing tasks. It was a significant milestone in robotics and AI, demonstrating the integration of perception, planning, and control.

► **Expert Systems—** "In the 1970s and 1980s, expert systems gained prominence. These AI systems used rule-based reasoning to mimic the expertise of human specialists in specific domains. Examples include MYCIN, an expert system for diagnosing infectious diseases, and DENDRAL, which analyzes chemical compounds.

► **Deep Blue—** "In 1997, IBM's Deep Blue defeated world chess champion Garry Kasparov in a six-game match. Deep Blue was a supercomputer that utilized advanced search algorithms and evaluation functions to assess chess positions. This victory marked a significant milestone in AI's ability to tackle complex strategic games."[5]

AI is not a new thing. What is new is that it's now mainstream both from consumer use (think apps and websites) to speed of development (think WordPress and cloud computing).

This is not the first time that technology has shifted from being accessible to a few to being available to many. In Dr. Peter Diamandis' book *Abundance: The Future is Better Than You Think*, he describes innovation as something that creates abundance where there was once scarcity. Invention is different from innovation. Apple did not invent the smartphone, but they did innovate it by turning it into a mainstream device that we can no longer do without.

In 1998, during the dot-com bubble, my company had a client who paid the going rate of millions of dollars to create an e-commerce website that had less functionality than what Shopify provides for $90 per month today. Technology changes fast and what was once hard becomes easy over time.

The innovation of advanced technology has always been surrounded by both optimism and skepticism. With this notion that innovation will "change everything," we are often initially reticent (thanks to our human instinctual fear of change) before we can dig into the opportunity.

As we noted in the preface, when spreadsheet software was first introduced to PCs, many predicted the end of human accounting departments. Why would we need accountants when the software does everything? As far as I know, the human-powered accounting industry is doing just fine. The opportunity enabled accounting firms to start consulting on implementing that technology. No corporate enterprise believed that their accounting departments were a business differentiator. Software that could optimize the back office efficiently and accurately was aggressively adopted. Companies like SAP and Oracle shifted from using accounting programs to ERP (Enterprise Resource Planning) software. The accounting companies most trusted by CFOs were Arthur Andersen, KPMG, and Deloitte. After all, they helped with accounting *and* they knew their business well. Arthur Andersen jumped on the opportunity and created Andersen Consulting, now known as Accenture.

What if organizations utilized AI to handle the mundane operations of the "salt mine" so they could focus on creating the intangibles? Organizations that bolster their positive intangibles will see their enterprise value reflected by market interest instead of an accountant who ran a discounted cash flow analysis to determine valuation. The quantifiable arguments of value are far less interesting than the qualitative aspects of intangibles. This book aims to provide background on how business has evolved to this point, where we are headed with this current surge of AI and other technologies, and how to navigate the transitions.

CHAPTER 2

WHAT IS AI, ACTUALLY?

In order to understand the effect that AI will have on companies and organizations going forward, it's important to understand what types of AI technologies are already in the marketplace. AI is more prevalent than you think, and in many cases, the success of the technology depends on the human operator.

In this section, the definitions are excerpts from ChatGPT, and then I will provide context on its applications in several fields.

► **Machine Learning—** "Machine learning is a subset of AI that focuses on enabling computers to learn and make predictions or decisions without being explicitly programmed. It involves training algorithms on large datasets, allowing them to learn patterns and relationships to make accurate predictions or classifications.

► **Deep Learning—** "Deep learning is a specific type of machine learning that uses artificial neural networks with multiple layers to process and analyze complex data. It has succeeded in image and speech recognition, natural language processing, and computer vision.

► **Natural Language Processing (NLP)—** "Natural language processing enables computers to understand, interpret, and generate human language. It involves speech recognition, language translation, sentiment analysis, and chatbots that can understand and respond to human queries or commands.

► **Computer Vision—** "Computer vision focuses on enabling computers to understand and interpret visual information from images or videos. It involves object recognition, image classification, object tracking, and image generation.

► **Robotics and Automation—** "AI technologies are also used in robotics and automation systems. Robots can be equipped with AI algorithms to perceive and understand their environment, make decisions, and perform tasks autonomously.

► **Expert Systems—** "Expert systems are AI technologies that embody specialized knowledge and rules to solve specific problems or make decisions in a particular domain. These systems rely on a knowledge base and an inference engine to reason and provide recommendations or solutions.

► **Reinforcement Learning—** "Reinforcement learning is a type of machine learning where an AI agent learns to interact with an environment and improve its performance by receiving feedback in the form of rewards or penalties. It is often used in applications such as autonomous driving, game playing, and control systems.

► **Generative Models—** "Generative models are AI models that can generate new data samples that resemble the training data. They are used in tasks such as image synthesis, text generation, and creating realistic simulations.

"These are just some of the main types of AI technologies, and there are many other specific techniques and approaches within each category. AI technologies continue to evolve and advance, offering exciting possibilities for various domains and applications."[6]

While there's a tendency to think of AI as this technological phenomenon that arrived in 2021, we can cite examples of AI-specific devices and capabilities that have been around for decades. Chatbots, for better or worse, have been utilized as customer service tools for quite some time. On a related note, as anyone who's called the main switchboard of a large telecommunications company or insurance provider can attest, they make a compelling argument for human ingenuity over the limitations of automated responses and computer interfaces.

There are countless examples of robotics and automation in the marketplace today. A well-known example, iRobot's Roomba, was created in 2002. While the Roomba is undoubtedly convenient—the menial household chore becomes far less of a drag when outsourced to a robot—it is far from a perfect substitute for personally vacuuming every last crevice, corner, and floorboard of your house. In 2022, twenty years after the creation of the

Roomba, the U.S. market size of the vacuum cleaner industry was over $5.65 billion, and less than one-quarter of the vacuums sold were robotic.[7]

It is undeniable that AI is poised to impact all sectors of the global economy. Let's look at some examples of industries where AI technology is already being used, and where it's having the most influence.

Agriculture

One good example of AI being used in agriculture is in the monitoring and management of crops. AI can help farmers make better decisions about when to water, fertilize, and protect their crops from pests and diseases.

For instance, AI-powered systems can use sensors to collect data on soil moisture, temperature, and plant health. The AI analyzes this data and provides useful insights to farmers. It can tell them precisely when and how much water or fertilizer their crops need, helping them optimize resource usage and reduce waste.

Additionally, AI can help detect and diagnose diseases or pests affecting crops. By analyzing images or data collected from drones or cameras installed in fields, AI algorithms can identify signs of disease or pest damage early on. This allows farmers to take prompt action and prevent the spread of problems, potentially saving their crops from significant damage.

Overall, AI in agriculture assists farmers by providing them with valuable information and recommendations. It helps them make informed decisions and manage their crops more efficiently, improving yields and sustainability.[8]

Medicine

AI is being used in healthcare to help doctors and medical professionals make more accurate diagnoses and treatment decisions.

For example, imagine a patient with a suspicious skin lesion. AI can assist in this scenario by analyzing images of the lesion and comparing them with a vast database of known skin conditions. By recognizing patterns and similarities, the AI system can provide suggestions or even identify the specific condition the lesion might indicate, such as melanoma or a benign mole. This helps doctors make a more informed diagnosis and decide on the appropriate treatment plan.

AI can also analyze medical images like X-rays, CT scans, and MRIs. By training with a large number of images and learning from expert radiologists, AI algorithms can help spot abnormalities or potential signs of diseases that human eyes might miss. For instance, in lung cancer detection, AI models can analyze lung scans and highlight areas that may contain cancerous nodules, assisting radiologists in the screening process.

AI can also analyze patients' electronic health records, lab results, and genetic information, to identify patterns and risk factors associated with certain diseases. By considering a multitude of variables, AI systems can generate predictions about a patient's likelihood of developing a specific condition or estimate the effectiveness of different treatment options for personalized care.

Healthcare professionals can benefit from additional insights and improve patient outcomes by leveraging AI's pattern recognition and data analysis capabilities.[9]

Transportation

AI is making self-driving cars possible by helping them see, recognize, and make decisions on the road.

For example, self-driving cars use cameras to see what's happening around them. AI algorithms analyze the camera images to identify objects like cars, people, and traffic signs. This helps the car know when to stop, turn, change lanes, etc.

AI also helps self-driving cars learn from many examples. They study data from human drivers, who show them the correct actions for myriad situations. By learning from this data, these cars can make smart choices when driving on their own.

Self-driving cars get even better by using a learning technique called reinforcement learning. They practice driving and get rewards for making good decisions. This helps them improve their driving skills over time.

A well-known project in self-driving cars is Waymo, by Alphabet Inc. (formerly Google). Waymo has developed advanced technology and has driven millions of miles in real-world situations. They use AI and sensors to navigate cities and have shown that self-driving cars can be safe and helpful.

AI is helping self-driving cars "see" the world, learn from the best examples, and make appropriate decisions. This technology is advancing quickly and has the potential to transform transportation in the future.[10]

CHAPTER 3

WHAT IS AN INTANGIBLE ORGANIZATION?

An Intangible Organization has a set of characteristics that make it unique, scalable, and highly defensible against competitors. Intangible Organizations can thrive in ever-changing economic environments due to great adaptability and foresight.

In an Intangible Organization, the value and competitive advantage stem from assets such as intellectual property, patents, brands, software, data, networks, and human capital. These organizations have often operated in the technology, research and development, consulting, education, software development, and creative service industries. However, it is now possible to be an Intangible Organization in any industry and to create entirely new types of businesses that may not yet exist.

Intangible Organizations are able to adapt quickly to changing market conditions, due to their emphasis on innovation and knowledge creation and their reliance on highly skilled professionals. They often have decentralized structures that promote collaboration, flexible work arrangements, and use technology to facilitate employee communication and knowledge sharing.

In the arts and other fields where creativity reigns supreme, efficiency and accuracy are rarely the attributes that fuel artists' success and popularity. These artists and creative organizations are valued by their "it" factor. While talent identification may come early, monetizing that talent and its value comes much later. How do we create organizations whose value can be scaled and monetized? Can a company become the next organizational Taylor Swift?

In other words, how can a company evolve so that its metrics (i.e., Swift's music) are almost secondary to its branding and storytelling (Swift's popularity as a cultural icon that drives concert ticket sales, social media following, Netflix documentary views, etc.)? Swift's superpower is her connection with her fans. In the same way, companies that intimately understand their customers and market their products with this understanding in mind will differentiate themselves.

The negative narrative surrounding AI and other technology is leaving many people intellectually listless, believing AI will replace their jobs. Instead of resisting or denying this new reality, we should focus on what makes us special and how we can leverage our uniquely human intangible assets.

AI was created in the 1950s, so it is by no means a new technology. What is new is that AI technology is now accessible to the masses and businesses in all industries, specifically ones that depend on droves of people to operate. AI technology is on the smartphones of CEOs worldwide, and they are experiencing the power firsthand of how this will change the game. In my innovation workshops for executives, over 90% of attendees say they are not questioning if AI will disrupt how they do business but when and how it will disrupt how they do business. Let's focus on the opportunities rather than the fears that AI may present.

INCREASING OPERATIONAL EFFICIENCY

For any innovation to find space and get its footing, there needs to be some creative destruction in the process.

Creative destruction is a concept introduced by economist Joseph Schumpeter. It is the process of innovation and technological change that leads to the destruction of existing business models and associated economics, jobs, and industries. This destruction creates space for the invention of new business models, creating long-term economic growth and progress. While there are short-term pains, creative destruction creates long-term gains. While I am sure the job losses impacted by the demise of Blockbuster were painful, video streaming services like Netflix have not only benefited the consumer but also the creators of video content. The limitations of video production and shelf space were a limiting factor in producing more movie and TV series content.

In a future where AI can execute most of the operational work in a company, intangibles will play a much more significant role in shaping one's career. While AI can perform routine tasks efficiently, unique human qualities and intangible skills will become increasingly more valuable and sought after. The general thinking that AI will eliminate people's jobs may be true in some cases, but it will also lead to the creation of new jobs as eliminating tasks will add time, space, and energy for inventing new opportunities. The goal is not to focus on the replacement aspects of AI, but to focus on its enhancement opportunities.

"Intangibles" refer to qualities that are difficult to measure or quantify but are significantly responsible for our personal and professional success. These intangible skills include creativity, emotional intelligence, critical thinking, adaptability, leadership, communication, and problem-solving abilities. These skills are not easily replicated by AI or automation, making them increasingly more valuable in a world where technology handles the more repetitive and transactional aspects of work.

Let's look at how intangibles play a role in one's career in an AI-driven operational landscape.

Creativity

As we have seen the advent of creator economies leveraging mass distribution via podcasts, blogs, YouTube, etc., we have also seen the advent of the curator economy. These companies do not create content but repurpose, reformat, and redistribute it.

It's similar to what happens in a corporate setting with the flurry of CC'ed and BCC'ed email forwards. I have a simple rule about opening forwarded email. If you are going to send me something that you didn't originally write, give me your informed opinion on the content and why you think I should read it. Otherwise, the forwarded email goes right into my Deleted Items folder. The idea here is that you should add value to everything you touch.

AI is much better at curating than humans are, but we mere mortals can do what AI cannot. Create.

As a venture capitalist, I meet with fifteen to twenty company founders weekly. They have ideas ranging from interesting to "moon shots." Most of their "good ideas" aren't good fits for me because for every $2 million I invest, I need to see a path for a $20 million return. Some businesses simply don't get big enough, quickly enough to justify this type of VC funding.

The saying that "ideas are great, but execution is what matters" has never been more true. However, the gap between idea and execution continues to shrink. Ignore the infomercial scam artists who say, "Got a great idea? Call us." Then say they want to patent your idea but, as very few people have patentable ideas, they will just be taking your money. These scammers will happily take all forms of credit cards and would probably cash your social security check for you.

Intellectual property does matter when considering a patent, but it isn't always applicable. The number of ethical and unethical intermediaries between Idea and Execution can be so overwhelming, that people either give up too soon, or worse spend hundreds of thousands of dollars to get nowhere.

While AI is currently hogging the spotlight, many other advanced technologies deserve our attention. 3D printing, cloud computing, frictionless e-commerce, etc., may look like roadies next to the AI rockstar, but they matter now more than ever.

When I was in college, I was hustling every way I could to make a few bucks. You name it, I was trying it. I was picking up and delivering the dry cleaning for fraternity and sorority houses, importing and selling towels and clothes from India, and I once purchased and loaded a shipping container of antique furniture in India and sent it to the US to sell. It would have been a great idea had the furniture been actual antiques, and not just old. I also read *The Wall Street Journal* almost every day and in 1991, I came across an article about how the EPA was going to have a new requirement for gas station pumps to have a pressure release system.

Naively, I said to myself, "You are an engineer and you can figure this out." I was naive not because of a lack of engineering acumen but rather because I did not consider that the gas pump manufacturers had known about this requirement for years and probably already had a solution. I sketched a rudimentary design, did some CAD work, and spoke to an attorney.

The attorney wanted a $10,000 retainer to start doing an IP search. He said it was a great idea; I mean, he wouldn't have asked for $10,000 if it were a bad idea, right? He then explained that I would need to hire a real engineer, develop prototypes, talk to mass manufacturing companies, and do more things that required exponentially more money than I had. My guess was that I needed around $1 million to execute my idea. In reality, I probably would have needed $10 million, but I didn't know any better at the time. As a twenty-one-year-old, $10,000 might as well have been $1 million, so I abandoned the idea.

How would that scenario play out today? I could come up with the idea in the morning at a coffee shop, search patent databases to see if it existed, use a free web-based CAD system to design it, iterate and push to a 3D printer service and have some prototypes the next day, hop on LinkedIn to contact decision-makers, show a prototype over a Zoom call and…you get the picture. The gap between my idea and its execution is so much smaller than it was in 1991. Thinking about AI, in the near future, you may be able to describe your product to an AI device while on the treadmill, approve it, send it to a 3D printer, etc. You'll get your thirty minutes of cardio and have just invented a new product. The inventor's mindset is the future. Write those ideas down now and then execute them tomorrow.

The human imagination, when unleashed, can develop the most creative inventions. These inventions often get scaled down and walked back to align with what was physically feasible at the time. Amazing ideas that get relegated to the mean. Technology like AI and robotics have the potential to deliver ideas without compromising them.

AI can assist in generating ideas and patterns, but it is the invaluable human creativity that drives breakthrough innovations. The ability to think outside the box, connect disparate ideas, and develop novel solutions will be increasingly important for career advancement. This is happening across the board. I could have had AI write this book. It would have been factually correct but a lot less interesting. Did I use AI to help me research? Absolutely. Did AI give me ideas or share my stories? Absolutely not. AI cannot feel, relate, and connect with others on an emotional level. Skills such as empathy, collaboration, and effective communication will be crucial in building strong relationships, leading teams, and navigating complex social dynamics at work.

Problem-Solving

AI systems excel at processing vast amounts of data and providing insights, but it is human judgment, critical thinking, and the ability to approach complex problems from multiple angles that will be in demand. These skills involve analyzing information, making sound decisions, and adapting strategies as necessary.

AI technology evolves rapidly and is therefore constantly changing the job market. The ability to adapt to new technologies, quickly learn new skills, and embrace change will be essential for career growth. Being open to continuous learning and seeking new opportunities for personal and professional development will be crucial.

Leadership

While AI can provide data-driven insights, human leaders inspire and motivate teams, set strategic directions, and make difficult decisions. Leadership skills, the ability to inspire, influence, and drive change, will be highly valued as organizations rely on them to guide them through complex challenges.

AI systems operate based on algorithms and data without a sense of ethics or moral judgment, though ethical frameworks and decision trees can be created over time. One of the most common discussions regarding AI and ethics involves the self-driving car. If a self-driving car is forced to choose between running over an old lady on one side, running over a child on the other side, or continuing straight into a barrier and endangering the passenger, how would it decide?

I am not sure what city these self-driving experiments are happening in, but no thanks. I don't care how good their food truck scene is.

A human would struggle with this decision as well, and it is less of an ethics question and more of a human question. It falls into the "it depends" category. If this construct continued to appear, AI would have a sample set of decisions and outcomes to judge against. Human professionals will play a critical role in applying ethical principles and making moral judgments in areas where AI falls short. These areas include addressing biases, ensuring fairness, and making decisions that consider the broader societal impact.

In an AI-dominated operational landscape, individuals who possess and continue to develop these intangible skills will have a competitive advantage in their careers. Individuals who emphasize and cultivate these qualities will thrive in roles that require creativity, complex decision-making, emotional intelligence, adaptability, and ethical considerations—areas where human capabilities complement and cannot be replaced by AI.

CHAPTER 4

CONTEXTUALIZING THE TOOLS OF THE FUTURE

The day my parents bought me an Atari 2600, I was pretty sure I would no longer need to pour quarters into Defender or Pac-Man at the video arcade. It was an amazing gift for a twelve-year-old kid from Stone Mountain, Georgia, and way better than that stupid *Pong* game that my brothers played. I still wasted a lot of quarters at the arcade and spent a lot of time indoors.

That said, avoiding playing outdoors was partly fueled by the city of Atlanta being on edge in 1981 as we waited for the child-targeting serial killer to be caught. Every evening during that time, the local channels ran a well-meaning but rather haunting recorded PSA that asked, "It's 9 p.m., do you know where your children are?"

Like many kids and adults, early exposure to computing was driven by fun, especially for this self-proclaimed nerd. I was genetically engineered by a civil engineer father and a software programmer mother, and intellectually nurtured by a public school that managed to procure an Apple Computer and only let the kids in the "gifted" program use it. I really never had a chance to be a cool kid.

When I was thirteen, my dad felt like his suburban Atlanta kids weren't Indian enough, so he decided to move back to India and take me with him. My older siblings had a choice in the matter and chose to stay in Atlanta. It was a dream of his to go back to India and give back to the community. My mom stayed in the U.S. to keep earning her salary, while my dad set up shop abroad. He started a civil engineering firm, bought an IBM PC (because some of his professor friends told him it was the future), and decided that I was best equipped to "figure it out," which I did to the best of my ability.

I created and automated marketing letters for him and printed them on a dot matrix printer. There was prestige in those perforated edges of the letters. He figured out that all the calculations he was doing on his HP calculator could be done on the PC, which became my next task. I must have succeeded well enough since he didn't complain.

He agreed with his professor friends that personal computers were the future and that they would change *everything*. Our IBM PC was placed in the middle of the office so that clients and employees knew that we were innovative. Though, for me, it meant that I couldn't inconspicuously play video games on it, which was its only redeeming purpose. My dad employed a small team of engineers that he was convinced had no idea what they were doing. He turned his conference room into a classroom to teach them what he was sure they didn't know and, as an added bonus, reminded them of how dumb they were. I never had to look up "generational trauma" in the dictionary, as my dad once snapped the eraser off my pencil and told me that erasers were for kids who didn't know how to do math. I sought refuge in the pixelated (and now ancient-looking) games on that giant white box of a PC. I am sure the engineers were curious about what I was doing on that box in the middle of the office. If they were, they were probably too scared to ask.

The lesson here is that if you scare the shit out of your people, they are unlikely to be innovative or even curious about much of anything at work. This is important to keep in mind in the context of the discussion of AI. "The future" cannot simply be forced down employees' throats. When exploring innovations and new technologies, it is critical to contextualize how to use these new tools and how they can help us create additional value and be more productive.

My father wanted to display the PC in the middle of his office simply to show that we were innovative, but he never took the time to show the engineers how to use this new machine. Unsurprisingly, using the computer as a scare tactic didn't yield positive results as they probably thought they were going to be replaced by the white box in the middle of the boss's office.

We must view AI through the lens that if individuals are told that AI will disrupt everything (including possibly their jobs and livelihoods), they will not view these advances positively. On the other hand, if we demonstrate how humans have many skills that AI doesn't possess, like creativity, problem-solving, and leadership—while showing how AI can fill operational gaps—then together we will be much closer to properly harnessing the potential of AI, and creating real value.

CHAPTER 5

AUTOMATING THE MUNDANE

I have yet to meet anyone who loves every aspect of their job. People tend to hate certain facets of their job because either they are not highly skilled at it or because it no longer makes economic sense for them to do those particular tasks "below their pay grade."

In the former scenario, hiring someone who can do the job you don't enjoy or don't excel at is fairly easy. The broad assumption is that they know more than you do, and that's why you recruited them. The only conflict is reflected in costs and value. This is mostly due to the ego of the hirer who asks, "can I really hire someone who makes more than I do?" The answer is yes, if the skill set they bring has a market value greater than your own.

The latter example of hiring someone to do a job function that you are fully capable of executing but have since moved on from, is not just emotionally challenging but practically challenging. The work has become mundane and you don't want to do it, but at the same time, you are great at it. The bar is set high for the person who enters this role and can be extremely frustrating for both parties. If you are a patient teacher, you may enjoy ramping someone up to the expert level that you have attained. It makes sense until that person gets promoted to a better job (or quits) and you have to train their new replacement.

If you don't have a teacher mentality, then you can invest your energy in utilizing AI for this task rather than spending it on more training cycles. The unpredictable and varying learning curves involved in training employees are highly inefficient and costly to organizations. You can utilize AI once and it will continuously get better. Self-driving cars are another great example of this. We start as terrible drivers at age sixteen. Over time, we learn and gain experience that makes us better until we hit the sunset years of our lives, and become terrible drivers again. Computers, on the other hand, don't follow a bell-curve learning path; they follow an exponential learning path. Netflix has a great way of suggesting movies I may like that I would've never found on my own.

There has always been a construct in business of focusing on your core competencies and supplementing the rest by building, buying, or partnering. Partnering is the easiest. If it doesn't make sense to have an in-house legal team, you partner with a law firm for support. Buying is more complex due to sourcing, capital requirements, and many other aspects. When GE needed to scale up their business process outsourcing, they acquired their business process outsourcing (BPO) vendor. They had the capital and resources to turn their costly partner into an acquired profit center. Building a new competency is the most difficult. There are so many unknowns that it costs twice as much as you budget for and takes twice as long as you plan for.

Why is building so hard? You have to assemble a team that has likely never worked together, has never done these building tasks before, and has the pressure of being under the microscope, to boot. There is also a scaling problem. The overhead and infrastructure required to build have historically been expensive and not meant to replace a job function but rather an entire functional unit. AI and other technologies are changing this.

TAKING CARE OF "SALT MINE" ACTIVITIES

Like many of us during the pandemic, I found myself grounded and working at home with continuous Zoom calls. We had a newborn and my in-laws moved in to help out. Every morning, I would grab a cup of coffee and tell the crew I was "off to the salt mines" as I trudged down the stairs to my basement office. A year later, I mentioned to my therapist that I hoped that AI would relieve me from working in the salt mine. Like any classic therapist, he asked, "What does the salt mine mean to you?" Ugh. I explained that the "salt mine" for me was all the tedious parts of my job that I hated doing and had yet to find the right person or software to do it for me.

While much of the discussion around AI has centered on jobs being replaced due to automation, there is a significant benefit that is not being mentioned. AI can make us more productive because it can automate the mundane tasks that we either dislike doing or are not worth our time. The upside of AI is that it can take over the tedious minutiae of any job, leaving us to focus more on the creative or higher-level tasks that add real value.

You have two solutions to cease doing the tedious tasks that make up your job. Either find an AI app to handle those tasks or, for your mental health, find a new job. I look forward to a future where I can spend less (or none) of my time on my "salt mine" activities.

DEMING MAY BE ROLLING IN HIS GRAVE

Let's look at automation in another way. W. Edwards Deming was an American statistician, engineer, and management consultant widely regarded as one of the leading figures in the field of quality management and process improvement. He is best known for his work in Japan after World War II, where he played a significant role in the country's economic recovery and transformation into a global manufacturing powerhouse.

Deming outlined fourteen key principles for managers to improve organizational effectiveness. These points include creating constancy of purpose, adopting a long-term view, embracing quality improvement, promoting employee education and training, and fostering teamwork. Deming popularized the Plan-Do-Check-Act (PDCA) cycle as a framework for continuous improvement. The PDCA cycle involves planning, implementing, measuring the results, and adjusting based on the findings. It is a systematic approach to problem-solving and driving ongoing improvement.

Deming stressed the importance of using statistical methods to understand and improve processes. Statistical process control (SPC) involves monitoring and analyzing process data to identify and address variations and improve quality and performance.

Deming also focused on the employee's intangible value to the business, as well as efficiency. He believed that fear in the workplace stifles creativity, innovation, and improvement. He emphasized the need to create a supportive environment where employees feel safe to voice their ideas, take risks, and contribute to continuous improvement efforts. He emphasized the importance of understanding and meeting customer needs and expectations and advocated for a shift from focusing on short-term profits to long-term customer satisfaction and loyalty.

The key takeaway from Deming's work by engineers and managers was creating a highly repeatable process and training of personnel. While concepts like Total Quality Management (TQM) focus on highly structured processes and workflows to achieve higher levels of quality, once it is structured and mapped, it becomes focused on training people on these processes. The idea is that a factory worker doesn't need to understand how a car works as long as they can follow the ten simple steps in their process and follow the color-coded process placards.

Many industrial engineering disciples of Deming want to train people to be robots. Now that we have robots, we don't need Deming—or possibly even industrial engineers. AI and robots, on their own, can improve processing in increments that humans cannot imagine. This may be in the form of adding or subtracting microseconds of movement to an ever-shifting supply chain. AI and robots can adapt as the Plan-Do-Check-Act is now embedded in software code. Continuous improvement is an algorithm and instead of training people to behave like robots through extensive training and management, we will just invest in and acquire robots.

The focus on efficiency was only half of Deming's philosophy. The other half was incredibly prescient as we think about evaluating businesses based on intangibles, rather than management and operations. If managers measure an employee's value to the company solely on short-term output without considering their intrinsic motivation, it is ultimately counterproductive to the health of the company. Workers need to not only understand processes, they need to care about how the car works.

In a *Harvard Business Review* article from 2016 titled "The Management Thinker We Never Should Have Forgotten," Joshua Macht wrote, "Many management thinkers have built upon Deming's philosophy, yet his core message seems lost to time. He argues that businesses destroy more value than they create when they focus on short-term results, traditional incentives, and performance rankings. His main point is that leaders must build deep trust among workers and managers, which emanates from a strong purpose and shared values. It seems logical enough—and more important than ever. So how is it that more businesses don't heed his message today?"[11]

As organizations begin to integrate and implement AI tools, it will become easier to increase productivity by automating the mundane tasks that are monotonous and time-intensive. If we look narrowly at Deming's teachings, it's easy to view workers as robots, judging them based on their ability to carry out the PDCA. In this instance, the industrial engineers who ruthlessly measure value solely based on efficiency are missing the point.

As automation becomes more prevalent, viewing employees for their expressly human traits is even more critical. A company with a strong sense of purpose and shared values, that makes hiring decisions based on how workers will complement those values, will be the most set up for success.

Simply put, processes can be automated—values can't.

CHAPTER 6

THE MONOLITHIC CORPORATION

Monolithic Corporations have been portrayed in TV and films as companies that are secretly (or not so secretly) working against humanity in some way due to the greed of a leader and rarely the greed of the organization. The stories often involve an elaborate chase scene where the hero is pursued (in a Keystone Cops manner) by nondescript security guards as he tries to dismantle the evil corporation. There is also the satirical portrayal of Monolithic Corporations like *Office Space*, *Silicon Valley*, and the like. The infamous "TPS Report" in *Office Space* became an embedded reference in corporate America as a way to make fun of itself. A hyperfocus on optimization has made corporate America resemble this. The classic, art imitating life scenario.

Since the Industrial Revolution, we have spent decades seeking ways to optimize and create self-sustaining businesses. We have celebrated the breakthroughs that icons like Ray Kroc (the innovator behind McDonald's) brought to restaurants for scaling and standardizing the fast food industry. Before Kroc, it was just the food industry. Then, the Chicago native innovated the entire system of managing costs and driving volume, as well as creating amazing training programs to de-skill the work, management, and KPIs of success. Essentially, Kroc's philosophy was to create money-printing machines and create amazing experiences.

His goal was to deliver a quality product. The reality is that quality means "consistency," not necessarily "good."

Marketing strategies convinced consumers that McDonald's burgers tasted the same (or better) than homemade ones and that price mattered most. Count pre-teen Eddie Murphy as one of the convinced customers. In his standup comedy film *Raw*, he recounts a story about being so upset that his mom making a "McDonald's burger" at home, instead of taking him to McDonald's. I am sure his mom's burger was better.

Throughout this book, I will refer to and compare the Monolith Co. with the Intangible Organization. This is the fork in the road that AI is driving. In this example, the McDonald's quarter pounder is Monolith Co., and the Intangible Organization is your mom's homemade version. While I have a personal preference for one of them, they are both an option.

SUCCESS IN ONE KEY VALUE DISCIPLINE

I routinely tell young entrepreneurs that *The Discipline of Market Leaders*, by Michael Treacy and Fred Wiersema, is required reading. In this following section, I also used ChatGPT to help summarize the book's key points.

"The book focuses on the concept of value disciplines and provides a framework for achieving a competitive advantage in the marketplace. The authors argue that companies can succeed by excelling in one of these three value disciplines: Operational Excellence, Product Leadership, or Customer Intimacy. Each discipline requires a distinct set of strategies and operational practices.

► **Operational Excellence—** "This discipline focuses on delivering products or services at a lower cost and with greater efficiency than competitors. Companies that excel in operational excellence strive to minimize costs, streamline processes, and maximize productivity. They often offer reliable, standardized products or services at competitive prices.

► **Product Leadership—** "This discipline centers on innovation and creating superior products or services that stand out in the market. Companies that pursue product leadership invest heavily in research and development, constantly strive to improve their offerings, and aim to be at the forefront of their industries in terms of technology, design, or features.

► **Customer Intimacy—** "This discipline revolves around building deep, long-lasting relationships with customers. Companies that prioritize customer intimacy focus on understanding their customers' needs and preferences, tailoring their products or services to meet individual requirements, and providing exceptional customer service. They aim to create a strong bond and loyalty with their customers."[12]

With AI, operational efficiency and product leadership are no longer distinctive traits. Meanwhile, customer intimacy, which has become a lost art, may be back.

While my preference is not to build a monolithic corporation, it is more than a viable option in an AI-driven world. AI can drive key efficiencies across every functional organization within a corporation. It will provide companies with an opportunity to grow their existing businesses in an even more profitable manner. In the near term, companies will be able to experience profitability like they have never experienced before. It will also significantly reduce the company's headcount within the management layers. The long-term effect of staff reductions would also diminish new ideation while keeping the focus on profitability.

The Innovator's Dilemma by Clayton Christensen was published in 1997, yet most of the core principles hold true. While corporations have made the book required reading, very few have executed the core concepts. The monolithic corporation is highly focused on its existing business models and believes that the benefit of AI is to drive costs out of its existing business models. AI adoption is not trivial, yet many companies erroneously believe they can adopt AI for cost reduction and innovation, which is very unlikely.

We can consider the following as traits of a monolithic corporation.

- ▶ **Reticence to new ideas—** They have highly established processes, structures, and routines that create resistance to change. The defensive position is stronger than the offensive position regarding any new idea.

- ▶ **Preference for established processes—** Their employees are comfortable with the status quo and are resistant to new processes and approaches. They focus on potential risks instead of opportunities for ideas.

- ▶ **Slow-moving industries are the most likely to be upended—** Laggard industries are the most risk-averse and also have the largest opportunity. Tech-savvy companies can benefit from change, but the laggards have the most to gain. They also are the ones most likely to be disrupted.

- ▶ **More focused on processes than consumers—** The measurement of everything, except for what matters, seems to be a key trait in monolithic corporations. Executives and management are very focused on developing KPIs, reports, and other dashboards. They are focused on managing the metrics and not the business or the customers.

► **Focus on short-term metrics—** Whether it is due to the ownership structure (private, public, employee-owned, or single owner), short-term performance goals and financial pressures prioritize immediate results over long-term innovation investments.

FEWER PEOPLE, MORE REVENUE, MORE CASH, MORE INVESTABLE DOLLARS

Notorious B.I.G's phrase "Mo Money Mo Problems" actually sums it up quite succinctly. More money means more stuff, more stuff means more to take care of, and more to take care of means more people to take care of said stuff. People are the challenge. Historically, all companies have a dependency on hiring people to scale their revenue. The most challenging is when people are the cogs.

Even technology companies with higher revenue/headcount ratios still need people. In this case, their main need, talented people, is also a scarcity. A good metric to track here is Annual Revenue/Total Headcount, aka Revenue per Employee (RPE). This metric is an efficiency ratio that estimates the average revenue a full-time employee generates. At the time of writing this, these were the tallies for the following companies:

- ► AECOM $269,000/full-time employee (FTE)
- ► Walmart $303,000/FTE
- ► Delta Air Lines $572,000/FTE
- ► Microsoft $1 million/FTE
- ► Google $1.6 million/FTE

In all these cases, it's actually "mo money mo people," but the economies of scale are very different. A traditional industry, like retail and construction, needs to employ people to maintain revenue and add more to grow it. Technology companies could cut research and development (R&D) investments and maintain their revenue—to a certain degree—with a lot fewer people.

In business, we have created so much vocabulary to describe our business's size, health, or success. These words matter and the words are changing. The culture of how we describe business metrics will continue to move towards financial metrics because the others will be rendered more or less meaningless.

Whenever there is a major shift, building new vocabulary and throwing out the old terminology becomes the foundation of being able to adapt to change. When company leadership in monolithic corporations are asked, "How big is your company?" they give responses like the following:

- *Going for our Series A*
- *$500,000 Annual Recurring Revenue (ARR)*
- *360 lawyers*
- *2,000 engineering professionals*
- *$5 billion per year in construction*
- *$1 billion in Assets Under Management (AUM)*
- *1.5 million square feet of space*
- *$1 billion in real estate (RE) transactions*
- *1,200 stores*

Rarely do we use $100 million in annual revenue with a 40% gross margin and $10 million in EBITDA or, even better, cash profit/headcount.

The vocabulary we use to describe our businesses (except for the last couple) are all indicators of what we think success looks like for our business. The last ones from the list are all that matter.

In contrast, the Intangible Organization looks at metrics more simply. It is very tangible when it comes to revenue, cash flow, and wealth creation. The building blocks of an advanced brand strategy, delivery models, and financial analysis that are powered by technology drive the new organization. The math of the business is tangible, but the enterprise value is intangible. I am not saying that in an AI-driven future, there will be no metrics. I am saying that they will not be a proxy for the metrics that actually matter. A business without metrics is a ship without a compass. For an Intangible Organization, $Annual Revenue/Head, $Annual Net Cash/Head, and Balance Sheet Growth are important. In 2009, after a brutal economic cycle, I sold my startup RCMS to NYSE-traded ARC. In the simplest of terms, ARC was a publicly traded printing services company. Their idea of technology at the time was to create internal efficiency and make it easy for their clients to print construction drawings, signage, and the like. Their acquisition of my technology company was driven to incorporate a digital strategy. Their core business (unlike mine) had weathered the Great Financial Crisis (GFC), and they thought I could bring a new digital strategy to their business.

They are a global business, and their idea was to acquire my company and bolt it onto their sales and marketing. While it appeared that they had avoided the GFC with masterful cost management, in fact, they had only delayed the impact. Within months of being acquired, they intensely scrutinized my business and tried to contain costs. I was asked, "Do you know how many sheets of paper we need to print to cover your losses?"

The executive team at ARC was, and continue to be, a group of very shrewd operators. It was a big cultural misfit. My company, RCMS, lived in the realm of the intangible and ARC was part of the world of Monolith. Eventually, I left to join Gehry Technologies. I felt like I had learned my lesson and that by joining Gehry Technologies, a firm started by world-renowned architect Frank Gehry, I would get to live in an intangible world. After all, clients paid Frank Gehry a lot of money to design buildings that were works of art. However, I was on the technology side. The technology side, in many ways, was developed to support the intangible architecture side of the business. It was an amazing year of working on great projects, but it wasn't a cultural fit. After all, I'm a technologist and civil engineer—not an architect. In an AI-driven future, a decision will need to be made. Create a strategy to build an Intangible Organization or a Monolithic Corporation.

The parallels to *The Matrix* seem relevant here. The 1999 film directed by the Wachowskis depicts a future where machines have overtaken society.

Morpheus, looking to wrest humanity back into human control, gives Neo, the protagonist, a choice between comfortable ignorance and reality.[13]

"This is your last chance. After this, there is no turning back. You take the blue pill—the story ends, and you wake up in your bed and believe whatever you want to believe. You take the red pill—you stay in Wonderland, and I show you how deep the rabbit hole goes." —Morpheus in *The Matrix* (1999).

Just as Neo did, we have a choice between facing reality and remaining in a state of comfortable ignorance.

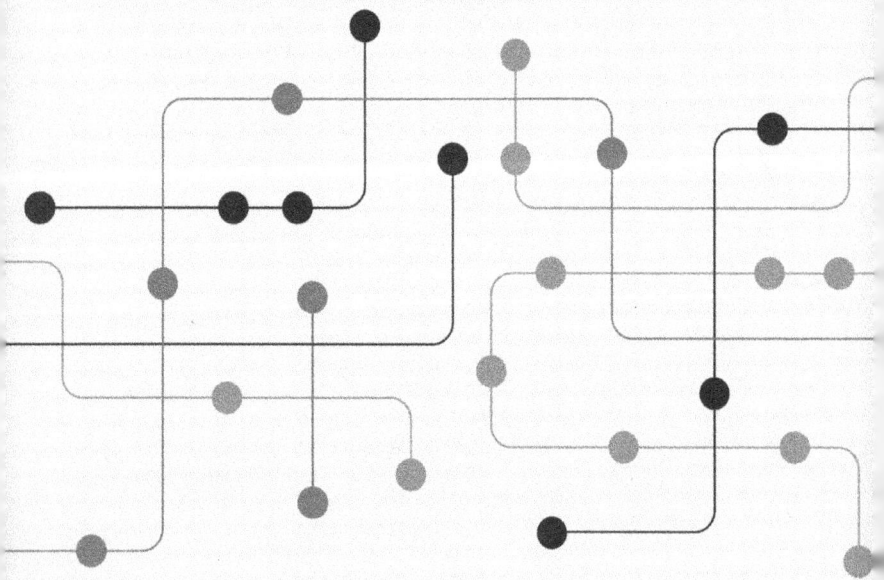

PART TWO

HOW TO THRIVE IN THE FUTURE

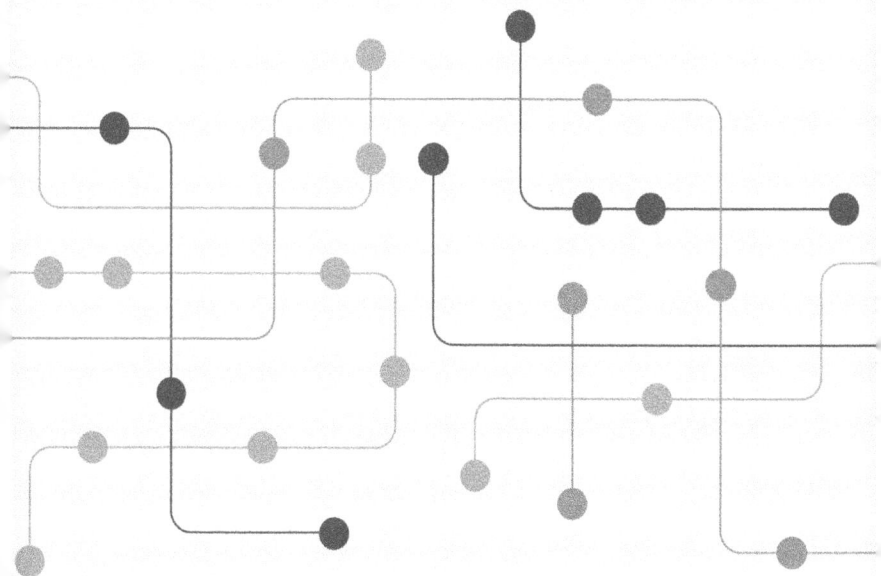

CHAPTER 7

LEADERSHIP THAT MOVES MARKETS, NOT PEOPLE

The term leadership is not just ingrained in business these days, but in every aspect of life. So much of leadership in the corporate world has been utilized to create a culture of growth. Leadership has been picked up as an intangible beyond management. Many believe that good management practices are a good defense, while leadership is about offense. The countless leadership books, classes, workshops, and organizations are designed to develop great leaders.

If machines don't need managers and we all become leaders—and we have all gone through the same training—then how are we different?

Leading a team, a department, or a company may not be enough. A leader with intangible characteristics often appears in the political arena, but how about in the business arena? The intangible leader now must look at how they move and influence markets, not just people—a higher calling and vision to move entire industries.

One of my favorite books is *Start with Why* by Simon Sinek. The book gives insights into how great leaders and organizations inspire others by starting with a clear understanding of their *why*. He argues that most organizations tend to focus on the *what* and *how* of their products or services (features, functions, and processes), without giving enough attention to the fundamental question of *why* they exist. The *why* represents the core purpose, cause, or belief that drives an individual or organization. It goes beyond making money or achieving goals and taps into the deeper motivations that inspire action. I love his concept of the "Golden Circle," which consists of three concentric circles: *why* in the center, surrounded by *how*, and finally, *what* on the outermost ring. Sinek explains that successful leaders and organizations start with the *why* and then move outward to the *how* and *what*. By starting with the *why*, they are able to effectively communicate their purpose and values, which in turn attracts like-minded individuals and loyal customers. Sinek also believes that "value" lies with the intangibles.

COMPARING LEADERSHIP STYLES AT APPLE

For my generation, we can't ignore companies like Apple. In addition to his signature black mock turtleneck, Jobs had intangibles. He shaped entire markets and consumer thinking about the future.

In fact, Apple's creation of the iPhone fundamentally changed not just how we build companies, but how we live. The invention of the smartphone gave every consumer a computer with Internet access in their pocket. With a GPS at our fingertips, we could seamlessly navigate through neighborhoods we've never been. We suddenly had libraries of thousands of songs on our devices. We had the ability to search and order anything, regardless of whether we were near a traditional desktop computer.

With the advent of the smartphone, an entire industry of mobile apps was born in the late 2000s and early 2010s. Social media companies like Snapchat and Instagram sprang up, allowing users to post photos and videos of themselves instantly, while others like Facebook and Twitter pivoted to take advantage of mobile traffic. Chase, Bank of America, Capital One, and others focused on mobile banking experiences—and now going in person to a branch to deposit a check is a relic of the past. Every quick-service restaurant, from McDonald's to Starbucks to Chick-fil-A, has an interface allowing you to place an order even if you're miles from the nearest location. It is impossible to overstate how much Jobs' first smartphone touched every industry on the globe.

While leading Apple, Jobs introduced numerous other products that became indispensable for consumers, including the iMac, iPod, and iPad. In addition to creating sleek devices that every customer wants, perhaps no leader was tied as closely to their brand as Jobs was to Apple. This is even more astounding when you consider the fact that he wasn't even a programmer. He was the embodiment of a leader with intangibles.

In contrast, when was the last time that you heard someone quote Tim Cook, Jobs' successor? Tim Cook is a very different leader. He is a skilled manager of business and people, but he lacks the aura and charisma of his predecessor.

On the one hand, this might seem like a counter-intuitive example. Cook may lack Jobs' signature black turtleneck, but Apple still has a market cap of $3 trillion and is the most valuable company in the world. And yes, Cook has made some decisions that have indeed changed the industry. The app tracking transparency update to iOS 14.5 in 2021 forced users to opt-in to allow apps to track their activity. Apple's privacy changes have significantly hindered Facebook's ad business, as well as other e-commerce brands.[14]

With this in mind, though, when was the last time Apple invented a new product that fundamentally changed the industry? Apple has continued to optimize its product line under Cook, but it hasn't produced anything that spurred innovation in the same way previous products did.

We can look at their contrasting leadership styles in a different way. Jobs' leadership was about offense—creating new devices that formed entirely new products. Cook's management practices have resulted in good defense. The concept that industry leadership moves markets rather than individuals refers to the significant influence and impact that leading companies and organizations have on an entire industry or market sector. Instead of individual consumers driving market trends, it is the actions, innovations, and strategies of industry leaders that shape the direction and dynamics of the market as a whole. This idea underscores the power and responsibility that lies with those at the forefront of an industry. Here are some key points to consider when exploring this concept in more depth.

Setting the Direction for an Entire Market

Industry leaders are often the trendsetters and pioneers within their respective fields. Their actions and decisions can set the direction for the entire market. By introducing groundbreaking products, services, or business models, they can shape consumer preferences, market standards, and industry practices. Their innovations and strategies can create new market segments, disrupt existing ones, or redefine customer expectations. Industry leaders have a strong influence on the competitive landscape.

Competitors often look to leading companies for inspiration, market insights, and best practices. The success of industry leaders can spark intense competition and drive other players to emulate their strategies or differentiate themselves in response. The actions of industry leaders create a ripple effect, prompting competitors to adapt, innovate, and strive for market dominance.

Leading organizations play a pivotal role in establishing industry standards, regulations, and guidelines. They may actively participate in industry associations, collaborate with regulatory bodies, or set benchmarks through their practices. By taking the lead in shaping standards, industry leaders influence the rules of the game, affecting market entry barriers, product quality requirements, and compliance expectations.

Influencing Investor Confidence

Industry leaders' performance and reputation significantly impact investor confidence and capital allocation within the sector. When leading companies demonstrate strong financial results, strategic vision, and market dominance, they attract investor interest and capital inflows. This influx of resources can fuel further growth, innovation, and market expansion. Conversely, a decline in performance or loss of industry leadership can lead to investor skepticism and capital flight, affecting the overall market dynamics. Ecosystems are the lifeblood of an industry leader. They often have extensive networks of suppliers, partners, and collaborators. Their decisions and actions can affect the entire ecosystem of businesses that rely on or support their operations. For example, changes in procurement strategies, product requirements, or sustainability initiatives by industry leaders can influence the practices and investments of suppliers and partners, creating a cascading effect throughout the value chain.

The reputation and image of industry leaders have a direct impact on the perception of the entire sector.

Their actions and behavior can shape public opinion, consumer trust, and stakeholder perceptions. Industry leaders are often seen as representatives of their respective industries, and their actions can influence public sentiment, market confidence, and regulatory scrutiny.

Leading companies tend to attract top talent and serve as talent magnets within their industries. Their reputation for innovation, market leadership, and career opportunities makes them desirable employers. As a result, industry leaders have access to a pool of skilled professionals who can drive further growth, contribute to industry advancements, and fuel market expansion. Intangible leaders move markets, and the right people will follow these leaders.

IP IS NOT JUST PATENTS

Intellectual Property (IP) is a broad and multifaceted concept encompassing various legal rights and protections for intangible assets. While patents are one form of intellectual property, it is important to recognize that IP extends beyond patents and includes several other types of rights that play a crucial role in fostering innovation, creativity, and economic growth.

Trademarks are a vital component of IP. They provide exclusive rights to protect names, logos, symbols, or other distinctive elements that identify and distinguish goods or services in the marketplace. Trademarks enable companies to build brand recognition, establish customer trust, and prevent others from using similar marks that may cause confusion or dilute the brand's reputation.

Copyright protection safeguards original creative works such as literature, music, artistic creations, and software. It grants creators exclusive rights to reproduce, distribute, display, perform, and create derivative works from their creations. Copyrights encourage artistic expression, incentivize innovation in creative fields, and ensure that creators can control and profit from their works.

Trade secrets, on the other hand, are valuable confidential information that provides a competitive advantage to businesses. This can include formulas, manufacturing processes, customer lists, marketing strategies, and other proprietary information. Unlike patents, trade secrets do not require registration and can offer indefinite protection as long as they remain secret and proper measures are taken to maintain their confidentiality.

Industrial designs protect the visual aspects and aesthetic features of a product or object. They cover the ornamental or aesthetic elements that give a product a unique appearance. Industrial design rights prevent others from copying or imitating the visual design of a product, encouraging innovation in product aesthetics and promoting fair competition.

IP is not just about the legal construct of defensibility—many view IP defensibility as the "shield" for the business against competition. Sadly, there are so many ways to work around traditional IP. A strategy of an IP moat historically has leaned on legal defense. The moat must include a unique strategy beyond legal.

SUGAR WATER AS AN INTANGIBLE

After all, if Coke were just sugar water, it would not be a multi-billion dollar corporation. Since Dr. John Pemberton produced the syrup that became Coca-Cola and paired it with carbonated water in 1886[15], the beverage has always looked to do more than simply quench its customers' thirst. Coke's goal, and ultimate success as a product, is establishing a deep, emotional connection with consumers.

Coke is not selling a carbonated beverage. It offers a feeling of companionship and warmth shared at dinner tables, backyard barbecues, sporting events, concerts, etc. Over the last half-century, its list of endorsers has included not just stars but cultural icons like Elvis Presley, Marilyn Monroe, Taylor Swift, Elton John, LeBron James, and so many more. The list goes on.

Coca-Cola isn't selling a carbonated beverage. It is selling intangibles.

Beefy Allure

In the last chapter, we referenced how McDonald's scaled and standardized the fast-food industry. It convinced consumers that its quarter-pounder was better than a homemade patty. I'm willing to bet that Murphy is not alone in his assessment—there is an aura around the McDonald's hamburger.

How has McDonald's convinced so many consumers of this? Through decades-long marketing efforts that have established deep connections with consumers. There's nothing quite like that first bite after you've received your drive-thru order and drive away in your car. McDonald's has created an entire experience and emotional connection with its consumers that can't be measured solely through metrics.

I have seen so many founders spend so much time and money filing patents and other legal IP strategies without talking to a single customer. The future isn't legal IP—it is a combination of efforts much like Coke. Monolith Co. has lots of large legal bills to file and defend patents. An Intangible Company may have patents, but they are part of a bigger strategy and not THE strategy.

CHAPTER 8

UNLOCKING TALENT (THE KEY TO AN INTANGIBLE ENTERPRISE)

How do we begin with a mindset of creating an Intangible Organization? Let's think of an organization as a platform. Ethos, mission, leadership, and brand make up four core components of the new enterprise platform.

Organization Ethos

One of the core pillars is ethos. In the context of an organization, ethos refers to the overall character, values, and credibility of the organization. It represents the collective reputation, principles, and ethical standards that the organization upholds and is known for. Ethos in this context is about establishing the organization's identity, building trust with stakeholders, and shaping its public image.

Mission

An organization's ethos is often reflected in its mission statement and core values. These statements articulate the organization's purpose, goals, and the principles to which it adheres. The mission and values provide a framework for decision-making, guide the behavior of employees, and signal the organization's commitment to certain ideals or causes. Ethos is closely tied to an organization's culture, which encompasses the shared beliefs, norms, and behaviors within the organization. The culture influences how employees interact, make decisions, and carry out their work. A positive and ethical organizational culture fosters trust, integrity, and collaboration, which contributes to the organization's ethos. Just as core values are not singular, neither is the mission. While having a strong vision is critical, it must be supported by strong core values, as creating an Intangible Organization is built through a series of Missions.

Leadership

The actions and behaviors of organizational leaders play a crucial role in shaping the ethos. Leaders set the tone, establish expectations, and model the values and ethics of the organization. Their integrity, competence, and commitment to ethical conduct are essential in building trust and credibility both internally and externally.

Ethos is developed through the organization's relationships with its stakeholders, employees, customers, suppliers, shareholders, and the broader community. Building and maintaining positive relationships based on trust, transparency, and ethical practices contribute to the organization's reputation and ethos. Effective communication, responsible business practices, and engagement with stakeholders are key elements in this regard.

An organization's commitment to social responsibility and sustainability also contributes to its ethos. Engaging in environmentally friendly practices, supporting community initiatives, and addressing social issues can enhance the organization's reputation and demonstrate its ethical values.

Brand and Identity

The organization's brand and identity are closely tied to its ethos. The brand represents the organization's values, personality, and promise to stakeholders. Consistency in messaging, branding, and behavior across various touch points helps to establish a strong and trustworthy brand image, reinforcing the organization's ethos. Ethos is reflected in the organization's approach to ethical decision-making. Having clear policies and procedures, promoting ethical behavior, and creating a culture that values integrity and compliance are essential. Consistently making ethical choices and holding individuals accountable for their actions reinforces the organization's ethos.

THE RETHINKING OF CULTURE

Old thinking: "Hire for cultural fit." *The new hire shall conform to the established cultural norms of the organization.*

New thinking: "Hire for cultural contribution." *How will the new hire contribute to the culture in a specific way?*

In the early days of building startup teams, there was more focus on cultural fit than intangibles. Tactical skills are a given part of any job. Go out on the Internet and review any job description. They are full of requirements for capabilities with tools and experience with tasks. Rarely do they say anything about non-task or capability descriptions, as well as anything about the unique personal characteristics that the role would benefit from.

These job descriptions in many ways read as instructions for a robot or software. Early employees must be adaptable and being a deep expert focused on a single task will generally be replaced by AI or a robot. Over time, as the organization grows, the team grows. The next thing you know is that there are thirty people in the Accounts Payable department, managed by five people, and one of those thirty people is tasked with coding invoices to G&A (General and Administrative). Why? These practices are meant to create operational efficiency. I have seen large companies provide incentives to their AP teams to pay invoices on time.

In a future where all these activities will be executed by machines, middle management will be the first to go. As we have stated time and again, AI sets a level playing field for operational efficiency. If providing operational efficiency is an employee's primary source of value, that employee is a leading candidate for automation.

Middle management, however, also has the best opportunity to thrive in an AI-driven world. Their knowledge of what tactical resources do and how they align with the needs of executive management could be the linchpin of the organization's future. If we look back at Deming's philosophy from our chapter on "Automating the Mundane," tomorrow's middle managers should specialize in purpose and values, and focus less on process.

THE IMPORTANCE OF LEADING, NOT MANAGING

Jack Welch & The GE Way: Management Insights and Leadership Secrets of the Legendary CEO, written by Robert Slater in 1998, was considered a must-read and had nothing to do with technology. The book details how Welch improved the processes at GE by removing layers of management to create an efficient organization. A large part of GE's success was in its management training program. This was where all managers learned how to do things the "GE Way." Welch was well known for firing the bottom ten percent of performers. These would be the people who did not "fit" into the GE Way. Methods like the GE Way, normalizing thinking and process, were considered the most optimal way to create scale and enterprise value. At GE, every colleague had a small, wallet-size GE Values card. Jack Welch, the CEO of GE had the values inscribed and distributed to all GE employees, at every level of the company. This plastic card was quite special. It represented the aspiration of the company and served as a code of conduct for how we engaged with our colleagues, customers, and suppliers.

"There isn't a human being in GE that wouldn't have the Values Guide with them. In their wallet, in their purse. It means everything and we live it. And we remove people who don't have those values, even when they post great results." –Jack Welch

GE was such a juggernaut in its time, but where are they today? On June 26, 2018, General Electric's (GE) more than 100-year run on the Dow Jones Industrial Average (DJIA) came to an end. The last remaining original component of the index was gone. A company co-founded by Thomas Edison stopped being innovative. Not only is GE not on the DJIA, but there was a time in history when GE executives could name their price to leave and run another public company. They were considered to be the best managers in the world. This may have been true, but they were likely not great leaders. Famously, Robert Nardelli left GE to run Home Depot with one of the most lucrative compensation packages in history. He abruptly left Home Depot in less than a year as a failure.

As the world continues to change, organizations that outperform their peers will have to learn how to recruit and lead people with intangibles. They explicitly lead their teams and do not manage people, which generally means managing the metrics that people are supposed to execute. We manage machines, not people. When you look at a corporate job description, it reads as if they want to hire robots and not actual human beings. A job description rarely describes the human attributes needed for the position along with the requisite skill sets. They'll state, "must have five years of experience with Salesforce" instead of "must have the patience of a saint to work with a hyperactive and emotional ten-person sales team."

The role of AI in organizations has the potential to impact the need for middle management in several ways. While it's important to note that the effects of AI on middle management can vary depending on the specific industry, organization, and context, here are some general points to consider.

Efficiency

AI and automation technologies handle repetitive and routine tasks more efficiently and accurately than humans. This eliminates the need for middle managers who primarily oversee and coordinate such tasks. As AI continues to advance, many administrative and operational functions (traditionally performed by middle managers) can and will be automated, leading to a probable reduction in these management positions. AI-powered systems can provide organizations with real-time data and analytics, enabling decision-making at various levels of the organization. This direct access to information can empower frontline employees, allowing them to make more informed decisions and reducing the need for middle managers as intermediaries in the decision-making process.

With AI facilitating real-time data analysis and decision-making capabilities, organizations may shift towards a more decentralized decision-making model. This gives frontline employees and teams the ability to make decisions autonomously, reducing the need for middle managers to oversee every aspect of decision-making and approval processes. The future may consist of many but much smaller teams.

Democratization of Communication and Knowledge

AI-powered collaboration and communication tools can enhance connectivity and information sharing across the organization. These tools enable direct communication between employees, teams, and senior leaders, bypassing traditional hierarchical channels. In this scenario, middle managers may have less relevance as intermediaries in information flow and coordination. Instead of performing operational and administrative tasks, middle managers may be required to take on more strategic and transformational roles within the organization. AI can free up their time and allow them to focus on higher-level functions such as driving innovation, implementing organizational change, managing complex projects, and fostering employee development.

It's important to emphasize that while AI has the potential to disrupt the role of middle management, it does not necessarily mean the complete elimination of such positions. Organizations will likely need to redefine and reshape the role of middle management to adapt to the changing landscape. Middle managers may be expected to develop new skills and competencies, such as strategic thinking, leadership, and the ability to leverage AI technologies to drive organizational performance.

Overall, the impact of AI on middle management will depend on how organizations embrace and integrate AI technologies into their operations, the specific tasks and responsibilities of middle managers, and the strategic direction they choose to pursue in response to technological advancements. If middle managers are put in a position to carry out tasks that AI cannot do—like fostering the coaching and growth of their employees while carrying out the ethos and values of their organization—then middle managers have quite an integral role to play. If they are simply a vehicle for tasks that can be easily automated—like sending invoices on time—then organizations will view them as far more expendable.

FINDING NEW WAYS TO INNOVATE

Startups epitomize innovation. Whether it is technical innovation or business model innovation, there is so much that large companies can learn from startups. Over the last ten years, I slowly shifted from being an entrepreneur to an investor. I say slowly because I still have a lot of bad intellectual and emotional habits that draw me back to entrepreneurship. After my last exit, I had the opportunity to work and then briefly lead the Advanced Technology Development Center (ATDC) at Georgia Tech. Established in 1980, ATDC is one of the longest-running and most successful technology incubators in the country.

The primary mission of ATDC is to support the development and growth of technology-based startups and entrepreneurs. It provides a range of resources, programs, and services to help startups accelerate their growth and increase their chances of success. One of the things I learned and taught at ATDC was the processes and playbooks of creating a startup. Whether it's Eric Ries' *The Lean Startup* or Steve Blank's *The Startup Owner's Manual*, there has been a wave of startup books over the last decade. These playbooks are widely known and used by founders.

So as they all use the same playbooks, backed up with AI to build their pitch deck, code their minimum viable product, etc., how do you start to evaluate these startups? As we saw from our *Moneyball* example in the opening chapter, once everyone is working from the same playbook, there's no discernible advantage anymore. Companies have to innovate in other ways to differentiate. Once all the required criteria are met, it is important that my team and I spend time with the founder. Does this founder have the potential not just to execute the business objectives, but also to be attractive to the next VC to fund them? We spend time dissecting the intangibles of the founding team.

INNOVATING IN A RISK-AVERSE INDUSTRY

My experience in corporate innovation has mainly centered on the built environment. There are several reasons why innovation in the built environment can be slower compared to other industries. The built environment industry encompasses various stakeholders, including architects, engineers, contractors, suppliers, regulatory bodies, and clients. Collaboration and coordination among these diverse parties can be challenging, leading to slower decision-making processes and difficulty in implementing innovative ideas.

The built environment industry tends to be risk-averse due to the significant financial investments involved in construction projects. The fear of failure or unforeseen consequences leads to a preference for tried-and-tested methods and a reluctance to adopt new technologies or approaches. This risk aversion can hinder the adoption of innovative solutions. The built environment is subject to numerous regulations and codes that aim to ensure safety, quality, and compliance. While these regulations are necessary, they can sometimes act as barriers to innovation. The lengthy approval processes, strict compliance requirements, and resistance to change within regulatory frameworks can impede the adoption of innovative practices and technologies.

Construction projects often have long life cycles, involving multiple stages from design to completion. This extended timeline can make it challenging to incorporate new technologies or processes that emerge during the course of a project. Additionally, the turnover of professionals involved in different project phases can result in a loss of knowledge and hinder the continuity of innovation efforts.

The built environment industry has traditionally had limited standardization, with different regions or projects often having unique requirements or specifications. This lack of standardization can make it difficult to scale or replicate innovative solutions across different projects or regions. Furthermore, limited collaboration and knowledge sharing among industry stakeholders can slow down the spread of innovative ideas. The built environment industry is highly cost-sensitive. Clients and stakeholders often prioritize cost-efficiency and return on investment, which can discourage the adoption of innovative but potentially more expensive solutions. The focus on short-term financial considerations may hinder the long-term investment in innovation.

Despite these challenges, there is growing recognition of the need for innovation within the industry. Efforts are being made to overcome these barriers, such as increased collaboration, regulatory reforms, and the adoption of emerging technologies like building information modeling (BIM), prefabrication, and sustainability practices. As the benefits of innovation become more evident, and as the industry continues to evolve, there is potential for accelerated innovation in the built environment in the coming years.

THE CHALLENGES OF CORPORATE INNOVATION

Driving corporate innovation in laggard industries is difficult on so many levels. Corporate innovation can be challenging for several reasons. Here are some common factors that contribute to the difficulty of corporate innovation:

► **Reticence to Change**— Organizations often have established processes, structures, and routines that resist new ideas and processes. Employees may be comfortable with the status quo and resistant to adopting new ideas or approaches. Overcoming this resistance requires a cultural shift that embraces experimentation, risk-taking, and openness to new possibilities.

In the corporate environment, there is often a strong aversion to risk. Innovation inherently involves uncertainty and the potential for failure. Many companies prioritize stability and predictability, which can hinder the willingness to invest resources, time, and effort into innovative initiatives.

► **Preference for Immediate Results—** Corporate innovation often requires a long-term perspective, as the benefits and returns may not be immediately evident. However, many organizations have short-term performance goals and financial pressures that prioritize immediate results over long-term innovation investments. This focus on short-term gains can impede the allocation of resources and commitment to innovation.

► **Lack of Collaboration—** Large organizations often operate in silos, with departments or teams working independently and not collaborating effectively. Innovation often requires cross-functional collaboration and breaking down these silos.

However, organizational structures, hierarchies, and communication barriers can hinder the necessary collaboration and knowledge sharing. Innovation requires dedicated resources, including funding, talent, and time. However, these resources are often limited or in high demand for other operational priorities. The competition for resources can make it challenging to allocate sufficient support for innovative projects and initiatives.

Without a clear innovation strategy and well-defined processes, companies may struggle to effectively manage and support innovation initiatives. Lack of alignment with corporate goals, unclear decision-making processes, and inefficient resource allocation can hinder the progress of innovation efforts. Measuring the success and impact of innovation can be difficult. Traditional metrics and performance indicators may not effectively capture the value created by innovative initiatives.

This lack of clear measurement and evaluation criteria can make it challenging to justify continued investment in innovation. Disruptive ideas that challenge existing business models or practices may face resistance from stakeholders who fear the potential disruption to their roles or the status quo. Overcoming this resistance requires effective change management, communication, and stakeholder engagement. This resistance is an emotional construct and requires leadership to create comfort that failure will be a learning experience and not career limiting.

Identifying Intangibles

After my time in the built environment, I shifted to found Shadow Ventures and focused on investing in startups. These organizations have little-to-no metrics in the way of traditional valuation data to perform a DCF (Discounted Cash Flow). DCF determines the current value of a business purely based on future cash flow predictions. A discount is applied based on the risk/opportunity of those cash flows. There is big math, like TAM (Total Addressable Market), but once we determine that the market is big enough and the problem is painful enough, then we evaluate the founders. We are looking for the intangible traits that make a difference.

Spotting intangible traits, such as talent, potential, and valuable intangible qualities, can be considered an art in itself. It requires a combination of observation, intuition, experience, and a deep understanding of human behavior. Here are some key aspects that contribute to the art of spotting intangible traits:

Attention to Detail

The ability to keenly observe and pay attention to nuances is crucial in spotting intangible traits. This involves observing how individuals interact, their body language, communication style, problem-solving approaches, and their unique strengths and weaknesses. Understanding the context in which individuals operate is essential. This includes considering the industry, specific roles, and the environment in which the intangible traits are expected to be valuable.

Different traits may be more relevant in different contexts, so having a holistic understanding is important. Experience in a particular domain or field provides valuable insights and patterns that can aid in recognizing intangible traits. Experts who have spent significant time studying and evaluating individuals in a specific area are often better equipped to identify exceptional qualities and potential.

Intuition

Intuition, although difficult to quantify, plays a role in the art of spotting intangible traits. It involves trusting one's instincts and relying on a subconscious understanding of patterns and signals that may not be immediately apparent. Intuition can be honed through experience and exposure to various situations. Spotting intangible traits often involves recognizing patterns that indicate potential or exceptional qualities. This requires comparing and contrasting individuals, understanding commonalities among successful individuals, and identifying unique characteristics that set them apart.

Engaging in meaningful conversations and actively listening to individuals can provide valuable insights into their intangible traits. Effective communication allows for the exploration of their perspectives, motivations, aspirations, and thought processes, which can reveal hidden qualities.

The art of spotting intangible traits requires a willingness to learn and adapt. As technologies, industries, and societal dynamics evolve, the traits that are valued may change. Staying curious, embracing new knowledge, and being open to reevaluating assumptions are essential for staying attuned to emerging intangible qualities.

It's important to note that spotting intangible traits is not an exact science and can be influenced by personal biases and limitations. Therefore, a multidimensional and collaborative approach that involves diverse perspectives and feedback can enhance the accuracy and depth of identifying and assessing intangible traits.

We see this most commonly in the startup world. After all, what metrics and valuation data were there when twenty-something Mark Zuckerberg launched Facebook in his dorm room? There wasn't even a revenue model for the business. However, the investors identified the quirks as an advantage and bet on him despite the other evidence. That analysis of intangibles will be paramount for VCs in the age of AI.

CHAPTER 9

BUILDING A NEW COMPETENCY

After I left GT ATDC, I joined a startup called SoftWear Automation as their CEO. It was a company that was incubated at ATDC and founded by a couple of professors. The idea was fairly simple—robots making clothes.

For me, this was not exactly a sweet spot. But here's the thing. I wasn't an expert in robotics or apparel. So why was I in an industry I knew virtually nothing about? Because I'm sort of an expert at becoming an expert.

MASS CUSTOMIZE EVERYTHING

The company had raised $3.5 million, and I needed to get up to speed on the supply chain. The good news is that I know a lot about supply chains, but not because of a degree, my history as an entrepreneur, or because it's generally fascinating. I'm an expert because I know how to break down how companies operate to be successful.

First, I learned about the apparel industry supply chain:

1. **Consumers**
2. **Retailers**
3. **Brands**
4. **Contract Manufacturers**
5. **Resellers**
6. **Manufacturers**

Then, I pondered, what's unique about this supply chain? I figured out that contract manufacturers make a "widget" from the retailer's design. It goes like this: back in the old days, retailers made their own products. They'd come up with a cool product idea, design it in-house, figure out how to make it, then put it into production. They controlled the chain.

Now, there is virtually no institutional knowledge about how to make clothing. That was a key learning. For example, designers at a company like Lululemon don't actually know how their products are made. They design the products and then hand them off to contract manufacturers. That was important to realize because we thought that ultimately we would be selling robotic apparel systems to companies like Lululemon. Nope! Not even close.

I revisited what I knew: building technology. In that supply chain you'd think that architects own the supply chain, but they don't. The contractors are the ones who source materials and equipment to help achieve an architect's vision and keep projects on budget. So, in terms of apparel, now we have an influencer—the contract manufacturer—making the key decisions about the machinery that may be needed to produce garments... not the retailer or the brand.

The interesting part about this experience was the scrutiny I received that robots were going to take jobs from seamstresses. If you ever have the opportunity to visit an apparel manufacturer, whether stateside or in China, I don't think you would think that these are jobs that people actually want. Once you see what people get paid, then you understand why it makes no sense. The harsh reality also is that there is a ton of child labor employed in international markets. So, am I concerned about taking jobs away from twelve-year-olds?

We studied all types of markets, from household soft goods (towels, bath mats, etc.), to automotive, sports apparel, and even luxury apparel.

Here is what we learned about these markets:

One to Many

This is a highly replicated product with very few product variations. T-shirts come in six sizes, with the same design, and come in a ton of colors. This world doesn't need AI or robots. High repetition with high volume can be handled by dumb automation.

Many to One

This is the world of custom suits. A bunch of small tailors that make you a custom suit and charge $5,000 to $10,000. It turned out that for that price point, we could have probably built a robot to make custom suits. However, it turns out that rich people romanticize the idea of having a custom tailor and the overall experience.

If you have seen the movie *Kingsman*, it is a similar experience to the scene where Eggsy gets taken to the tailor, except for the cool spy toys.

Many to Many

This is the world of alteration shops. You see them in just about every shopping center. We looked at building a robot to serve that market as well. Imagine buying a pair of pants and having them hemmed while you wait. It turns out all these little shops work cheaply and retailers don't mind sending it out to a tailor and consumers are used to it.

Mass Customization

Mass customization is a new world. This is where we got very excited. What if a single cell of robots could make anything? Scan your body with an app, send it via the Internet, and pretty much like a Kinko's, pick it up at a local print shop. One robot cell that can make pants, shirts, t-shirts, underwear, etc. Softwear Automation is still at it and trying to figure it out.

This is a great example of how to think about AI relevance within an industry. In a many-to-one world driven by AI and robots, mass customization could create new markets, reduce all overseas shipping, and create a new type of product—a smaller domestic-based manufacturing base with a nimbleness to manufacture the latest fast-selling fashion trends.

Currently, the bestselling products sell out fast, creating aftermarket demand and the poor-selling products end up deeply discounted or worse, in a landfill. Did we ever think that athleisure would be a thing? Well, in full disclosure, I wrote most of this book in soccer pants and a T-shirt. It is not always a zero-sum game.

DEATH OF A CONSULTANT

What will the management consultants do? An organizational chart, also known as an org chart, is a visual representation of the structure and hierarchy of an organization. It illustrates the relationships, reporting lines, and divisions of authority within the organization. An organizational chart provides a clear overview of the different departments, positions, and roles within the organization.

An org chart displays the various positions or job titles within the organization. This includes top-level positions such as CEO, president, or managing director, as well as lower-level positions such as managers, supervisors, and individual contributors. The chart shows the reporting relationships between different positions within the organization. It indicates who reports to whom, illustrating the chain of command and the flow of authority.

Reporting lines depict the hierarchical structure and lines of communication within the organization. An org chart often represents different departments or divisions within the organization. It may include major functional areas such as finance, marketing, operations, human resources, sales, and customer service.

Each department is typically represented as a separate branch or section of the chart. Org charts often depict different levels of management within the organization. This includes senior management at the top, middle management in the middle layers, and lower-level management or supervisors closer to the operational roles.

Lines and connectors are used in an org chart to indicate relationships and connections between different positions. For example, a solid line connecting two positions indicates a direct reporting relationship, while a dotted line can represent a more indirect or advisory relationship. In addition to reporting lines, an org chart may depict functional relationships between different departments or roles.

I have some good friends in management consulting. They are good people. I think that historically they have had an important place in business. The sheer size of management consulting firms and the highly educated talent they attract have allowed them to be a central resource for research, knowledge, and insights that most companies could never employ nor retain.

In many ways, management consultants borrow ideas from client to client and then build data and strategies to execute these ideas. The problem is that ideas and research used to be difficult to come by for large corporations.

Management consultants are professionals who work with organizations to help them improve their performance, solve problems, and achieve their goals. They provide objective advice and expertise to assist companies in making strategic decisions, improving operations, and implementing change.

Consultants analyze an organization's current operations, processes, and strategies to identify areas for improvement. They conduct research, gather data, and assess the organization's strengths, weaknesses, opportunities, and threats. This analysis helps them understand the root causes of problems and identify potential solutions. Consultants assist organizations in developing or refining their business strategies. They work closely with management teams to define goals, identify market opportunities, and create plans to achieve competitive advantage. Consultants may conduct market research, analyze industry trends, and provide insights to help organizations make informed strategic decisions. Consultants help organizations streamline their operations and improve efficiency. They identify bottlenecks, inefficiencies, and areas where processes can be optimized. By employing tools and methodologies such as Lean Six Sigma, consultants redesign processes, develop new workflows, and implement best practices to enhance productivity and reduce costs. Consultants support organizations in implementing major changes, such as organizational restructuring, mergers and acquisitions, or technology implementations. They develop change management strategies, create communication plans, and help manage the transition to ensure a smooth adoption of new processes or systems. Consultants may also provide training and support to employees during the change process. Consultants assist organizations in assessing their financial performance and improving profitability. They analyze financial statements, develop financial models, and identify opportunities for cost reduction or revenue enhancement. Consultants may also help organizations with financial planning, budgeting, and forecasting.

Consultants work with organizations to improve their overall effectiveness and enhance employee performance. They assess organizational structure, culture, and leadership practices. Consultants may provide recommendations on talent management, employee engagement, performance measurement, and succession planning. Consultants often lead or support project initiatives within organizations.

They define project goals, develop project plans, allocate resources, and monitor progress. They ensure that projects are delivered on time, within budget, and meet the desired outcomes.

I was asked once if I would fund a startup led by a team of experienced management consultants. I answered with an aggressive "hell no." By the way, I was asked on a stage sitting next to two venture capitalists who had most recently been in management consulting with an audience of a few hundred people.

The audience member who asked the question was obviously bored and knew that I would take the bait. He asked for my logic. My logic was simple. Starting a company requires a lot of risk. Not just financial, but also reputational risk. It is a business of building and not overanalyzing the situation for months. Going fast and breaking things is the business. I was asked by a student with a solid B average from a great engineering school if they should go into management consulting. I told them that they weren't either good enough or interested in school based on their GPA.

Management consulting is a continuation of school. I like to describe them as personable academics. The path to building an Intangible Corporation does not go through management consulting or business school.

WHAT'S THE INCENTIVE FOR BEING A PUBLIC COMPANY?

A public company is a type of business that has issued shares of stock that are traded on a public stock exchange. This means anyone from the general public can buy and sell the company's shares. Public companies often have a large number of shareholders, which can include individual investors, institutional investors, and even other companies. Examples of well-known public companies include Apple, Microsoft, and Amazon.

On the other hand, a private company is not publicly traded on a stock exchange. Instead, its ownership is typically held by a smaller group of individuals or entities. Private companies can have a varying number of shareholders, ranging from just a few individuals to a group of investors or a single owner. Since private companies are not publicly traded, their shares are not available for purchase by the general public. Examples of private companies include Mars, Cargill, and Koch Industries.

The main differences between public and private companies lie in their ownership structure, level of financial transparency, and regulatory requirements. Public companies are subject to more stringent regulations and reporting requirements, such as regular financial disclosures, to ensure transparency for their shareholders and the investing public.

Private companies, on the other hand, have more flexibility in terms of reporting and decision-making, as they have fewer regulatory obligations.

Public companies often have greater access to capital through the sale of shares and can use their stock as a form of currency for acquisitions and employee compensation. Private companies, however, can maintain more control over their operations and strategic direction since they don't have to answer to a large number of shareholders. In a business model transition, being a public company is a massive disadvantage. The need to meet quarterly earnings is always the focus. This financial focus does not leave much space for a company to take risks or make tough decisions.

As we've seen, in this changing landscape, focusing on long-term growth rather than short-term wins is critical. The leaders of Monolith Co. though, do not have this luxury. Controlling strategic direction and taking calculated risks is much more difficult when you're answering to many stakeholders. In many ways, the goal of becoming a public company is a march to becoming a monolithic corporation.

CHAPTER 10

WHAT ARE THE TRAITS OF AN INTANGIBLE ORGANIZATION?

W hile there are great examples of companies that started as Intangible Organizations, many are unable to keep the key traits that made them intangible and transform into monolithic corporations. Apple, Google, and Microsoft are great examples of this. Most recently Meta (Facebook) was criticized for its strategy by shareholders, which led to a mass layoff of employees.

It is tough for public companies to stay intangible while being criticized every quarter by investors for every decision they make. As I studied companies both large and small, some key traits seemed to be common, both at startups and large corporations.

THE IMPORTANCE OF A DISTINCT AND DRIVEN BRAND

I love the podcast *Acquired* with Ben Gilbert and David Rosenthal and one of my favorite episodes involved LVMH. LVMH Moët Hennessy Louis Vuitton, more broadly known as LVMH, is a French multinational holding and conglomerate specializing in luxury goods. It's one of my favorite brands and not just because it was started by a fellow civil engineer. This company truly understands intangibles.

LVMH has about sixty subsidiaries that manage many well-known iconic brands. In addition to Louis Vuitton and Moët Hennessy, LVMH's portfolio includes Tiffany & Co., Christian Dior, Fendi, Givenchy, Marc Jacobs, Stella McCartney, Loewe, Loro Piana, Kenzo, Celine, Sephora, Princess Yachts, TAG Heuer, and Bulgari. LVMH epitomizes the creation of a massive enterprise value based on its luxury brand status.

Luxury brands are such an interesting world that live in the intangible universe of thought. Unlike a premium brand, which offers better quality or materials, services, or processes, a luxury brand is defined by perception and status.

Many business and self-help books stress the importance of building a personal brand. Consumer product companies do a fantastic job of creating brands. A brand is much more than a logo and words. Premium brands have great logos but their words are focused on the tangible and measurable. Characteristics like performance specs, quality awards, etc., give a great definition to a premium brand. Luxury brands, meanwhile, have great logos but are rarely seen in TV ads. Lamborghini has said that their customers don't watch TV. Instead, these brands sell a lifestyle. Their customers, like celebrities, perpetuate a brand that makes you feel just a little closer to a famous person.

We can learn a lot about intangibles from luxury brands. We assess their value based on association with other brands, a perception of scarcity, or how the brand changes our self-image. Brand matters now more than ever before.

Branding What's not Measurable

A brand is considered an intangible asset because it represents the perception, reputation, and overall value associated with a company, product, or service. It is a collective representation of various intangible elements that are not physically measurable.

Brands exist in the minds of consumers and stakeholders. It is shaped by their perceptions, experiences, and interactions with the company or its offerings. A brand's reputation, built through factors like quality, customer service, and messaging, resides in the minds of people. A strong brand often elicits emotions, feelings, and associations in consumers. These emotional connections are intangible and cannot be physically quantified. A brand's ability to inspire loyalty, trust, and positive sentiment is intangible and relies on the abstract aspects of human psychology.

Branding elements such as logos, slogans, and brand names can be legally protected as intellectual property. While these assets have physical manifestations (e.g., a logo design or trademark symbol), their value lies in the intangible recognition and association they create in the minds of consumers. Brands have inherent value that goes beyond the tangible assets of a company. The trust and credibility of a brand influence consumer purchasing decisions and perception of quality. This value is intangible and difficult to measure solely based on physical assets or financial metrics.

Brand Differentiation

Brands help differentiate a company or product from its competitors. They communicate unique value propositions, positioning, and personality traits. These intangible qualities contribute to the overall perception of a brand and its place in the market. Brands can evolve and adapt over time, responding to changing consumer preferences, market dynamics, and societal trends. This adaptability and longevity are intangible attributes that require ongoing management and strategic decision-making.

While a brand may have tangible elements associated with it (such as marketing materials or physical products), its true essence lies in intangible factors like perception, reputation, emotional connection, and value. These intangible aspects are crucial in shaping consumer behavior, influencing market positioning, and establishing long-term relationships with customers.

In the age of AI, branding will become paramount—even more so than it already is. As operational efficiency increases, the difference in what companies deliver in terms of materials and processes will become negligible. So what's left to differentiate companies? Perception, reputation, and emotional connection—the unmeasurable product specs.

Being brand-driven is crucial for businesses in today's competitive marketplace. It allows them to differentiate themselves from competitors and stand out in the minds of consumers. A strong brand cultivates customer loyalty, as it creates positive experiences and emotional connections. Trust and credibility are built through consistent delivery of brand promises, leading to long-term relationships with customers. Additionally, a well-established brand can command a price premium, increasing profitability. It also engages employees, providing a clear sense of purpose and attracting talented individuals. A strong brand facilitates expansion and diversification, leveraging its reputation and customer trust to explore new opportunities.

On the whole, being brand-driven enables businesses to establish a strong market presence, connect emotionally with customers, and achieve sustainable growth. Aligning your company vision to a brand ethos is essential for several reasons. It ensures consistent messaging, presenting a unified identity to your audience. This alignment builds authenticity, fostering trust and credibility with customers. It also engages employees, providing purpose and direction, and it helps differentiate your business, highlighting its unique qualities. It guides strategic decision-making and ensures initiatives are in line with your brand's direction. Most importantly, it creates an emotional connection with your target audience, leading to increased loyalty and advocacy.

Overall, aligning your company vision to a brand ethos is crucial for establishing a strong brand identity, building trust, engaging stakeholders, differentiating your business, and forming meaningful connections with your audience.

FEARLESS OF FAILURE

One of my favorite books is *The Psychology of Money* by Morgan Mousel. The core concept is "Anything that is huge, profitable, famous, or influential is the result of a tail event."

What does this mean? It means that failure is part of the process. You have probably also heard the term "fail fast." These two concepts are key characteristics of an intangible organization. Creating a portfolio of new products and services is important. They should deliver asymmetric business objectives. If not, they are a failure. Why? The risk has to be respectful of the reward.

If you are a $2 billion company, is a new product delivering $30 million per year with a 10% gross margin a failure that should be sold off or shut down? The simple test is how does it affect your key metrics of revenue per head and net profit per head?

The short answer? It doesn't. Failure is not always a zero outcome. Famously, if you're not first, then you're last. In Monolith Co., fearing failure is very different than failing at a startup. The career risk at Monolith Co. is years of relationship equity, pensions and retirements, and the comfort of working in an organization that you know well and have become comfortable with. The fallacy of the intrapreneur is that they are an entrepreneur but inside a corporation. The fear of failure is high, so the focus is on building products and services that are low risk and low return.

Overcoming the fear of change in this new AI-driven world can be quite daunting, but there are effective strategies that can help you navigate through it. One way to start is by recognizing the potential benefits that change can bring. For instance, a new project or initiative might provide an opportunity for you to develop new skills, gain valuable experience, or make meaningful contributions to your organization. By focusing on these positive outcomes, you can shift your perspective and view change as an exciting catalyst for personal and professional growth.

Seek information and understanding about the upcoming changes. For example, if your company is implementing a new software system, take the time to learn about its features, benefits, and how it will streamline processes. Understanding the reasons behind the change and how it aligns with the organization's goals can alleviate some of the uncertainty and fear associated with it.

Breaking down the change into smaller, manageable steps can also make it more approachable. Let's say your team is transitioning to a new project management methodology. Instead of feeling overwhelmed, focus on learning one aspect at a time, such as understanding the new terminology or familiarizing yourself with the updated workflow. By taking it step by step, you can build your confidence and gradually adapt to the changes.

Building a trusted network is crucial during times of change. Seek out colleagues, mentors, or friends who have experienced similar changes and can offer guidance and encouragement. They can share their own experiences, provide insights, and offer practical advice on how to navigate the transition successfully. Having a support system in place can help alleviate anxiety and provide a sense of reassurance.

Embracing continuous learning is another effective strategy. In the face of change, look for opportunities to expand your knowledge and skills. For example, if your company is adopting a new marketing strategy, consider enrolling in relevant courses or attending industry conferences. By proactively seeking out learning opportunities, you can equip yourself with the tools necessary to thrive in the changing professional landscape.

Visualization is a powerful technique that can help overcome the fear of change. Take a moment to imagine yourself successfully adapting to the change and excelling in the new environment. Visualize the positive outcomes, such as increased productivity, improved job satisfaction, or recognition for your contributions. By focusing on these positive mental images, you can build confidence and reduce anxiety associated with the change.

Taking action and embracing change is a crucial step in overcoming fear. Once you have gathered information, prepared yourself, and sought support, it's time to put your plans into motion. For example, if your company is restructuring departments, actively participate in team meetings, offer suggestions, and show enthusiasm for the new structure. By demonstrating your proactive attitude and adaptability, you can position yourself as a valuable asset during times of change.

Reflecting on past successes can also provide reassurance. Think back to previous instances when you successfully navigated change or overcame challenges. Recall how you adapted, learned, and grew from those experiences. Reminding yourself of your past achievements can boost your confidence and reinforce your ability to overcome obstacles in a professional setting.

By recognizing the potential benefits of change, seeking information and understanding, breaking it down into manageable steps, building a support network, embracing continuous learning, visualizing success, taking action, and reflecting on past successes, you can effectively overcome the fear of change in a professional setting. These strategies will enable you to navigate through transitions with confidence, adaptability, and a growth mindset.

USING LESSONS FROM YOGA TO STRETCH OUR PERSPECTIVE

I first started practicing yoga about twelve years ago. A minimum of a class a day. Those who know me are aware that I can be competitive. When I first started, I would ask the instructor to teach me "what that dude is doing." She politely told me to "stay on my mat." That was not the answer I wanted to hear. However, I listened and have been happy with my practice.

In the sport of innovation, the opposite is true. Don't stay on your mat (your silo). Be intellectually curious about everyone. Your direct competitors are obvious. What about a vendor or an adjacent business? This is also why ecosystems are so important. It helps you to stay in the know. Opportunities don't always exist in your field of view, so getting a different point of view is important.

Just as you are constantly trying to create value for your clients, how are your vendors creating value for you? So much of what we "outsource" to vendors is driven by either capability or capacity.

We often see this with software development. A lot of companies work with third-party software development firms to build their in-house software. These shops not only have talent that may not exist in the corporation but also have the ability to scale up or down with workload. In many cases, they may also have some internal best practices that you would have to build. However, an area that many are overlooking with AI is software development. In technology, there is a concept called the 10X

Developer. The concept is that 90% of all productive code is built by 10% of the software team, and there are lots of arguments around this concept.

However, with the advent of AI for software coding, early indicators are that these tools can take a mediocre software developer and make them a 10X developer. The scale effect of AI could drive a company to bring software development in-house and solve both the capability and capacity problems.

Besides cost, why do this? So much of the future is tied to the company's vision and mission, and it is terribly difficult to have an outside person to buy into this. In many ways, they are a mercenary for hire. Very capable, they do what they're told and don't ask too many questions. These people may not be the people that you want on the team to build an Intangible Organization.

EMBRACING UNCONVENTIONAL APPROACHES AND GROWTH MINDSETS

My wife recently shed light on the concept of working the grays versus the absolutes. Sometimes she has to remind me that I am an engineer and that most of my intellectual processing is binary. It appears that the gold is in the grays.

Working in gray areas refers to operating in the ambiguous or uncertain areas of a particular field or situation. It involves dealing with complex problems or situations that lack clear-cut solutions or guidelines. While these gray areas can be challenging and sometimes uncomfortable, there are several potential values in engaging with them.

Gray areas often require thinking outside the box and exploring unconventional approaches. By working in these spaces, you can foster creativity and innovative thinking. The absence of predefined solutions can lead to the discovery of new ideas, methods, and perspectives. Embracing gray areas helps develop your problem-solving skills. It encourages critical thinking, analysis, and the ability to navigate complex situations. By engaging with ambiguity, you can learn how to remain adaptable and find effective solutions even when faced with uncertainty.

Gray areas present opportunities for personal and professional growth. By stepping outside your comfort zone and grappling with uncertain situations, you can expand your knowledge, skills, and capabilities. It allows for continuous learning and development as you encounter new challenges and acquire valuable experience.

Many gray areas involve ethical dilemmas or situations with no clear moral guidelines. By confronting these challenges, you can develop your ethical decision-making abilities. It requires considering different perspectives, weighing potential consequences, and making thoughtful choices based on your values and principles. Taking ownership of gray areas can demonstrate leadership and initiative. When others shy away from uncertainty, stepping forward to tackle ambiguous problems can set you apart. It shows your willingness to take risks, assume responsibility, and find solutions even in the absence of clear direction.

How does this relate to an Intangible Organization? Working in gray areas cultivates adaptability and resilience. It helps you become comfortable with uncertainty and change as you learn to navigate uncharted territories. This flexibility enables you to better handle unforeseen circumstances and thrive in dynamic environments.

While engaging with gray areas can be challenging, the value lies in the growth, learning, and opportunities that emerge from embracing ambiguity. It fosters creativity, problem-solving skills, ethical decision-making, and personal development, ultimately positioning you to excel in complex and uncertain situations.

Let's look at a different but related example. In Carol Dweck's book, *Mindset: The New Psychology of Success,* she talks about a growth mindset versus a fixed mindset.

"People in a growth mindset don't just seek challenge, they thrive on it. The bigger the challenge the more they stretch."[16]

I think it is a great framing for something in the startup community that feels pretty obvious. I think with the advent of AI, robots, and innovation, a fixed mindset will drive the application of innovation that will drive a race to commoditization. A growth mindset will apply this innovation to develop yet-to-be-invented business models.

I have studied hundreds of entrepreneurs. The best entrepreneurs have a common trait: they understand the difference between friction and immovable objects. The fixed mindset has them using all their resources to push through an immovable object and running out of resources. With the growth mindset, founders have a better capability of making their next-best move when they hit a brick wall. In corporate America, the brick wall is the inertia of doing nothing.

I'm sorry to say that regardless of your business model, doing nothing is no longer an option.

CHAPTER 11

BUILDING THE TRAITS OF AN INTANGIBLE ORGANIZATION

While identifying the traits in Intangible Organizations can be helpful, it is probably even more helpful to learn how to develop them.

Again, Carol Dweck's insight from *Mindset: The New Psychology of Success* is relevant here. Dweck states, "Just because some people can do something with little or no training, it doesn't mean that others can't do it (and sometimes do it even better) with training. This is so important because many, many people with the fixed mindset think that someone's early performance tells you all you need to know about their talent and their future."[17]

STORYTELLING IS PARAMOUNT

One of the topics that I see frequently in the startup world is the focus on storytelling. Whether it is in a pitch deck, a pitch competition, or an investor pitch, startups are told to tell a great story.

Why? Because in the early days of a startup, there are no tangibles. No revenue, no funding, no real business plan, but rather an idea and a story. The story typically starts with how the founders observed a problem, how they started to do some market discovery work to understand the problem further, and how they recruited a couple of co-founders who were equally irritated by the problem, and hence they started their journey together. Then there are the stories of all the trials and tribulations of their work together and the market conversations that made them even more convinced of how important of a problem this was to solve and how huge it would be to solve it. Then the story continues about how they had their eureka moment of a solution. With such enthusiasm and momentum, they easily convinced some early customers, and now they are at my doorstep wanting to know if I want to join their rocket ship. They have room for me on this rocket ship but seats are going fast and I need to decide quickly.

This is why people love *Shark Tank*. They are great stories and the cast of characters (the Sharks) makes it even more entertaining. It is also a TV show whose revenue model is advertising. It's as much about investing as *The Bachelor* is about finding true love.

The challenge with storytelling in my experience has been an inverse relationship between great founders and great storytellers. Great founders solve real problems in novel ways. They are mostly bad storytellers and are frustrated when investors don't "get it." The others are fantastic storytellers with little substance. Great examples are Elizabeth Holmes, Sam Bankman-Fried, and Adam Neumann. Yes, many fraudsters are great storytellers. It turns out that my engineer brain doesn't love a great story—I prefer the introverted nerd founders who are bad storytellers.

However, storytelling is a vital skill in this new era of AI. Why? AI can't tell great stories or create them based on shared experiences. If a founder is going to simply provide a problem statement and a solution without a story, they will struggle. AI will write many good stories, but the human experience to date, and the accompanying journey, is something humans can only deliver. The nuances of obstacles and overcoming them with your one shot to go big is not an AI story. Do we really think AI can create a story where an entire stadium chants "Rudy! Rudy! Rudy!"?

An authentic story rooted in the people will always win. We love the stories of Edison, Bell, and others. Their challenges and friendly accidents are what connect people to your organization. The origin stories are what history is made of.

Startups have figured this out. It's time for the rest of businesses to learn storytelling, just as well as they know finance or engineering. If storytelling is left up to marketing, it won't have a soul. Intangible Organizations are not just about the talent teams but also their stories. An Intangible Organization is a book, and the stories of its people are the many chapters that fill it.

Does anyone really want to read a 10-K or 10-Q? Realistically, AI will generate those.

PERFECTING CUSTOMER EXPERIENCE

While the idea of a customer experience isn't new (Nordstrom and Ritz Carlton have been doing it for decades), it has had prominence with consumer brands. For a business-to-business enterprise, it has been more of a second thought. For an Intangible Organization, a well-designed customer experience is a core trait.

Make customers the top priority in everything you do by understanding their needs and making sure your business meets them. Treat customers as individuals and recognize that each customer is unique. Customize your interactions to their specific preferences and make them feel special. Being proactive, anticipating what customers might need, and reaching out to them with helpful information or recommendations. Showing them you're one step ahead, and you either know their business as well as they do or are interested in knowing their business better than they do.

Curiosity goes a long way to crafting a profound customer experience. Show empathy to understand and care about your customers' concerns. Listen to them and take action to address their issues. Provide a consistent experience across all touchpoints.

Your brand, messaging, and service quality should be the same wherever customers interact with you. A constant check against branding ethos, vision, and mission is a great way to maintain consistency. Keep track of how well you're doing. Use metrics and indicators to see how satisfied your customers are and identify areas for improvement. Most of all, keep having conversations with your clients. AI can't read body language and build rapport in a way to understand the truth of how the experience feels and how it can be improved.

MY FIRST BRUSH WITH ASYMMETRY

In 1997, I quit my civil engineering job and started building custom web application software for companies. We learned that while there was demand for creating websites and applications, there was even more demand for providing Internet service and connectivity. My partner and I decided to start providing Internet connectivity. It wasn't some great strategic move, but we just listened to our customers. This new business essentially meant that we had two businesses—a services business that built websites and software, and what essentially was a telecom business.

As we started designing our network, we bought routers and such to connect to our wholesale provider. We would buy a big Internet pipe, what is known as a T-1 (1.54MBS). This was 1997, so it was considered a lot of speed compared to the dial-up modems that went up to 28.8 kilobytes per second. The general idea was that we bought a T-1 for $1,500 per month and sold dial-up access for $29 per month. This meant that at any one time, we could have twenty-four people dial in at the same time.

The broad assumption was that we could sell more accounts than we had capacity since not everyone was going to be on the Internet at the same time. Also, this was a time when most people had one phone line at their home and they couldn't tie it up all evening by being on the Internet. It was also pretty common to get a busy signal and the customer would keep trying. So, we could sell a port that cost us about $100 per month (T-1 plus equipment lease, etc.) about ten times. About $300 per month for $100—it wasn't a bad business.

Then we noticed something. We had residential customers that mostly operated at night and weekends and the business customers didn't work nights or weekends. So, we started adding more business accounts. The business accounts also needed more help and had us set up their entire networks and security to the Internet. Now instead of ten accounts per port, we were running twenty accounts per port. Doubling our revenue with the same cost. Then we realized that as we kept adding more T-1s our ratios got even better. The laws of large numbers started to kick in. In the meantime, our software consulting services were also growing. The difference was that we had to hire people constantly, buy them computers, and give them office space (remember it was 1997). Our Internet service business had one person to handle tech support. Also, what would happen if we had a slow month? I still had to pay all these people.

I had a very large Fortune 1000 company call me and ask me to submit a proposal on a workforce automation project. It was a huge project for us. Our first six-figure project. I pitched hard and we were one of three finalists, along with Microsoft and IBM. Obviously, they were more expensive but I was asked in the final pitch, "Why should we give you the business, you are a small company?" My response was "I don't see the CEOs of Microsoft or IBM here. Must not be that important to them compared to me."

We won the deal. A day later I called the client and they faxed the contract over. Once I got it, I asked, "When can we get paid?" The client was taken aback. She said that we had to do the work, hit milestones, submit an invoice and they would pay us in thirty to sixty days. It was a hit to the gut.

I was twenty-six years old. I had zero idea that this was how they would work. They had plenty of cash—why wouldn't they pay us upfront? Didn't she realize that I had to hire people, buy computers, get office space, and buy desks (folding tables from The Home Depot)? Finally, she agreed to give us a small deposit to get us going. It was amazingly stressful.

Meanwhile, we received a call at our Internet services company. A large school system needed Internet services as soon as possible. We set up 500 accounts in a day (mostly data entry work) and sent them an invoice for a deposit and the first month's service. They wired money the next day. It didn't add any additional costs, it was just more money. I didn't know the word before, but this was asymmetry. I stopped trying to sell any consulting services ever again.

Why have people shied away from asymmetric business models? Historically the upfront capital required to establish them has been a barrier. This capital was required to hire people for their expertise or, in the case of manufacturing, necessary resources for R&D and building a factory. Consulting on the other hand, only requires some capability and a client. Get one client and you pay your mortgage. Get a few more and you get to buy a second home. Hire a few people to serve your clients, and the cash flow will be fantastic. With little to no capital outlay, a services company can be profitable on Day One.

Very few companies have been successful at developing scalable asymmetric business models due to the risk and capital outlay. However, one comes to mind.

AMAZON'S ASYMMETRY

While we can think of Amazon as this corporate monolith that has laid waste to smaller independent outfits near, far, and everywhere in between, throughout its history, it has ceaselessly innovated and continued to keep competitors on its heels. As Steve Wunker writes in an article for *Business Strategy Insider*, since Amazon's inception, it has utilized asymmetric business tendencies as a vehicle for disruption. We can take away key lessons from Amazon on the importance of both customer experience and asymmetric business models.

Amazon launched as an online bookstore slightly ahead of the dot-com frenzy, and it used a distinct business model to up-end a staid industry. By collecting payment from buyers well before it paid suppliers and by initially declining to carry any inventory itself, it could slash prices on popular titles and still make money through the "float" interest it earned on the money paid by users for purchases. Brick-and-mortar chains couldn't respond. Then, once Amazon had built up its own distribution centers, it could become a hub for users' e-commerce needs by selling and distributing products from rival merchants.[18]

While Amazon's business has drastically evolved since it started in Jeff Bezos' g arage in Bellevue, Washington, in 1994, one central tenet has been its intimate connection with and knowledge of its customers. Regardless of what consumers are buying—whether it's beef jerky, shark snuggies, convenient shipping, movie streaming, or cloud computing—Wunker points out that Amazon offers an "intuitive experience and provides good value" on everything it sells.[19]

That "good value" is lower pricing on banner goods while extracting revenue from the customer elsewhere. Amazon can afford to sell retail products at little-to-no markup because it charges customers $14.99 per month for the privilege of having anything under the sun shipped to them in two days or less. This is the perfect example of an asymmetric business model.

Moreover, Wunker also shows how Amazon's intimate connection with its customers allows it to quickly create and innovate other products. Once Amazon captured the connection of book readers by re-selling paperbacks and hardbacks, it introduced the Kindle in 2007. While the e-reader was a seemingly eccentric idea at the time, it allowed readers to store thousands of books on one device at the same time. Moreover, it offered readers the ability to consume any book or periodical instantaneously. "The direct relationship also gives Amazon the flexibility to adjust offerings rapidly,"

Wunker writes, "rather than having to rely on a tangle of business partners."[20]

Amazon's asymmetric business model is made possible by its intimate connection with its customers. That connection makes it possible to quickly monetize other goods while always attuned to the needs of the customer to improve and innovate its product line.

SQUEEZING MARKET LESSONS OUT OF A LEMONADE STAND

As we see from our Amazon example, so much of creating new products requires first understanding the markets. This requires analyzing the market problems. Existing businesses striving to re-invent and become Intangible Organizations often look at the markets they already serve. While this could seem like an obvious choice, leadership may say that they are fine with disrupting an existing service line, but for the rank and file, it is unnatural to cannibalize a service line that they know with an unproven technology solution.

Instead, consider adjacent markets to explore. An adjacent market could be driven by size, industry, or even a different customer in the same market. As an example, years ago as the cloud market was growing, SAP started a business called MySAP.

SAP targeted large multi-billion dollar companies. MySAP was their cloud-based, minimal-feature solution targeting smaller clients. A company with $500 million in annual revenue was their sweet spot.

The key thinking around "owning" markets is to create a monopoly. In fifth grade, we got an Apple computer at school. The computer was carted around from class to class in a wooden credenza with wheels. I was a nerd (which I always saw as a positive thing—I'm glad other people are starting to see it that way, too) so I was put in charge of it.

Out of the few games on the computer, the most popular one was *Lemonade Stand*. *Lemonade Stand* was a simple business simulation game. You bought cups, lemons, sugar, and ice cubes, and tried to determine your ideal recipe. In order to maximize your profits, you'd have to adjust for weather and other variables. I spent hours on the game, trying to run the best lemonade stand Lemonsville had ever seen.

Many entrepreneurs view an IPO as the height of success. My version of ultimate success would be the Department of Justice showing up and saying they need to break up my company because we're a monopoly. #goals.

As with many business situations, there is not one right answer. However, there is a right answer based on the situation. All startups should have a vision of becoming a monopoly. How do you create a monopoly when you're bootstrapping? Well, I've never done it, so take this with a grain of salt (or sugar in this case if we're talking about lemonade). But I think we can use the example of a lemonade stand to see how it's done.

Own Your House

Your first lemonade stand should be in your living room. Rent is free, and your materials are free too (borrow them from your mom's pantry). Sell to every friend and family member that comes to your house. Unless your siblings are going to compete with you, you'll be the number-one lemonade stand in your house within no time. It's a small market, but it's also a 100% gross margin business. If cash is the lifeblood of a business, gross margin is the heart. Iterate your recipe, iterate your process, and learn about your cost variables. Hoard your cash.

Own Your Street

Now that you own your house, it's time to own your street. Take that cash, build a stand, and maybe print some flyers. Hopefully, there will be limited competition. I doubt your mom will let you raid her kitchen anymore, so you'll likely have some COGS (Cost of Goods Sold). Printing those flyers costs money, and you may need some contract labor (your siblings) to pass them out. First lesson on staff: do you pay them hourly, on a fixed fee, per flyer, or per glass of lemonade sold?

Run your street until you understand the cost variables and revenue variables. Predictability reduces risk. Consider seasonality, correlation of demand to weather, and customer preferences. If your street is health-oriented, maybe you need to offer an organic version. Be careful, though—you can't afford to support too many SKUs. You're still not paying rent and you don't have any real overhead. And now you have a pile of cash. But you want a bigger pile.

Own Your Block

Moving from your street to your block means you need to optimize marketing and determine your service area. You'll be tracking costs and examining the efficacy of your marketing strategies. People may love your lemonade but they're not going to drive across town for a glass. Track customer activities (maybe using a simple CRM) and engage with customers to ask where they live and how you can improve.

It may also be time to get help in your operation. If customers are tired of waiting for you to serve and collect money, maybe your kid sister can run the cash register for you. That pile of cash has grown, and now it's been a year. You've done a good job of tracking your finances and metrics. You've built a brand, people on the block know about you, and you have a regular customer base. For every invested dollar, you're getting two dollars back. Not too shabby.

Own the Neighborhood

Here we grow again. Owning the neighborhood brings new costs and competitive forces. It's time to pay rent (no more working out of the house), it's time to go full time (not a side hustle anymore), and it turns out that people may want a smoothie instead of lemonade. The business just became real. You have a great handle on marketing and creating demand. The next marketing activity is to start understanding your competitive positioning.

The real focus will be on costs. With all the additional costs, watching every penny matters. Now is also a good time to seek out mentors: people who have been in the business and provide specific functional skill sets. As a rule of thumb, don't make this move until you have one year of expenses in the bank, including enough cash to pay yourself a reasonable salary. And guess what? A year or two later, you have a new pile of cash. It's the biggest you've seen in your life. Your business is real. Now can it be big? This is one of the most critical strategic decisions that needs to be made.

Next Moves

Once you own the neighborhood, where do you go from there? You follow the Theory of Adjacencies (not sure if I made this up or read it somewhere) to determine how to grow. The basis of the theory is to only make one-degree market moves. A two-degree market move is too much risk and generally fails.

First, plot your market on two axes (in this case, products versus geographic markets). A natural one-degree move would be to open a location in a neighborhood five miles away. An unnatural move would be to open a store an hour away. Another natural one-degree move is to start selling potato chips. An unnatural move would be to add a deli counter and make sandwiches. A third-degree move would be to add a deli counter to your store an hour away. You're bankrupt if you make that move.

Are you tired of talking about lemonade? Have I wrung everything out of this metaphor? I use this example because it applies to almost all businesses. In tech companies, geography may not matter, but your segmented markets will matter. You may not sell potato chips, but you are continually adding features.

Intangible Organizations operate like startups, and startups are always thinking about how they can own the market. In some cases, that means focusing less on short-term wins and more on long-term growth. In other instances, that may mean zeroing in on one-degree moves but maybe not third-degree moves. The key to re-invention is becoming hyper-focused on increasing your market share.

DISTRIBUTION MATTERS

At my annual Shadow Summit, I ask entrepreneur friends to come to inspire corporates and startups on their journey of how they built their unicorns. What I love most about these speakers is what I learn from them. We had my friend Tim Sheehan, founder of Greenlight Financial, speak at our Summit. Greenlight provides a credit card targeted at parents to teach their kids financial responsibility. It has been a huge success. I was fortunate to work with Tim when I was running ATDC. Tim is a fantastic human and a pragmatic builder. In his talk to us, he focused on how important distribution is to a startup for scale.

When Greenlight was approached by a major bank that wanted to be a distributor for the Greenlight card, the company had a unique choice. There was a risk that the bank could learn from Greenlight, but at the same time, it was an opportunity to have access to millions of prospects. Tim ultimately chose to partner with them.

His point was that as a startup with a unique solution, gaining market share was important and that distribution would give them a tremendous market. It was a huge success that ended up attracting many other financial institutions. A great focus on the first problem, first solution, and first market can be the difference between success and failure. This focus also helps understand who would be a great distribution partner because distribution matters.

Several years ago I was hired by a large beverage company to help them "think different." I spent the day understanding their product portfolio and noticed that they had several products that overlapped with customer personas. They didn't disagree and went on to share the numbers. It turned out that their bar for success was fairly low. It was also clear that their new products were popular for a short period, but didn't last. In many cases, they ended up buying smaller companies that were doing better at sustaining growth in these markets.

I looked at the executives and said "It's clear to me that your mass distribution keeps you from actually understanding the markets and buying preferences. It's a crutch for your innovation. I bet you guys could bottle cat piss and sell $100 million of it annually before the market signaled back to you that it was horrible."

While my delivery is not always ideal, the point landed. A startup has to work hard to earn distribution. Distributors have to want your product. A startup has no leverage other than being the best.

UNDERSTANDING MARKETS AND THEIR PROBLEMS

"Think different" was an advertising slogan used from 1997 to 2002 by Apple. In my advisory efforts with executives, this comes up pretty consistently. Jobs' famous term is now over twenty years old, but the fact that they are still stuck on it says so much. Monolithic companies develop strategic plans in a vacuum to "think different" with two main missing components. The first is that they do zero work to interrogate the market. Why do I call it an interrogation?

My recommendation to entrepreneurs embarking on customer discovery is to watch a few hours of *Law & Order*. The original, not the derivative work.

See, I've traveled about 150 days a year for the last ten years of my career. My Sunday afternoons "resting" on the sofa turned into binge-watching

Law & Order on TNT. I've since learned that it was not very good for my body and soul, and I stopped binging. But not before I realized that *Law & Order* has a lot in common with customer discovery.

CUSTOMER DISCOVERY LESSONS THROUGH AMERICA'S FAVORITE CRIME DRAMA

Customer discovery is the process of interrogating the market to understand the problem, whom it affects, and who is willing to pay to solve the problem. The keyword here is "who"—who is a person, not a company. All this work helps identify your ideal customer by establishing a common set of characteristics and traits. Simple, right? Not simple at all. But *Law & Order* can make it a little easier.

And here's why: you know that dance that Sam Waterston and Jerry Orbach always do? The one where Jerry believes that he's found the perp, and he wants Sam to prosecute. But Sam says he needs evidence. He can't prosecute off Jerry's hunch alone. In *every* episode, this dance happens. Well, that's about how customer discovery goes. In this case, I'm Sam Waterston and you're Jerry Orbach. So if you want to learn about customer discovery, let's dance.

Opening Scene

There's been a murder, and the detectives show up at the crime scene and look around for evidence. One of the detectives makes a quippy remark, quickly followed by the famous "dunt-dunt" *Law & Order* sound. Then the detectives take a look at the evidence and start interviewing people who knew the victim. In your case, the victim is the problem, and you need to figure out who has the problem. Your crime scene is your "problem statement," and early interviews help you create a solid "hypothesis."

The Hypothesis

On *Law & Order*, they always take a look at the spouse first. They are right sometimes. But they can't be right all the time (or they wouldn't have such a monster TV franchise). Maybe this time it's the mortuary owner, the cable installer, or the sister. As you iterate your hypothesis by interviewing people closest to the problem (the victim), you narrow it down to a few suspects to focus on. Then you pick them up for questioning.

The Interrogation

One at a time, you put each suspect in the "box." The first thing they do in the interrogation room is get the suspect comfortable. The detectives mirror the facts to date, but they always leave out key facts that have not been in the press.

In your case, you talk to the perp (sorry, the subject) about your journey and some key facts that you know. You throw in some names of the other people you plan on interviewing and add some other thoughts. You never put the cuffs on them, nor do you disclose your early hypothesis. You're still framing out your hypothesis. Then you let them go, and tell them to stay in town. Once you've completed all the interviews, you have a set of facts that exclude some people. Like the people who have an alibi or who lack a clear motive (you thought it was the spouse, but there was a prenup). You're now down to your final hypothesis, and you become very passionate about it. You have several facts but no proof.

The Ask

You go to Sam Waterston (or me), and you say you have all the evidence. You want me to go to the judge for an arrest warrant (the infamous investor pitch deck). In my case, you make the pitch, and you close by stating that you want a term sheet. Sam Waterston says, "You have some circumstantial evidence and a hunch, but you don't have proof."

Jerry complains, "What do I need to do... have them on camera shooting the victim?" and Sam replies, "That would be nice, but a confession would be just as good."

Jerry has arrested the perp (trying to sell them) but now has to release them. You tried to sell the person too soon instead of getting solid proof that their problem existed.

The Confession

In the next scene, Jerry has the perp back in the "box." His goal here is to get a confession. "Did you kill the victim?" The perp says no. In your case, you ask "If I can increase your profitability, would you be interested?" and the answer is yes. Both you and Jerry asked a question with a highly predictable answer. Please don't do it.

The Wrap-up

And here's where the actual customer discovery begins. The goal is to get a confession by asking a series of questions that catch the perp in a lie. In your case, your goal is to ask the right questions so that the person's tone and body language start to give you a better line of questions and indicate where to dig deeper.

This is why I always recommend customer discovery be performed in person. There's no silver bullet of questions here. I interrogate with negative questions that create an emotional effect. If I think there's a massive problem with the management of privacy (see my friends at OneTrust), then I would ask "Don't you think privacy laws are dumb? Such a waste of everyone's time…" In other words, get people emotionally engaged and make them take a position and defend themselves. It works on *Law & Order*, and it can work for you.

The second gap is a deep focus and love for the solution that they created. The pursuit of creating solutions that may or may not have a market can best be left to academics. The fog that a monolithic corporation knows best for its clients is always a huge miss. If you indeed understood your market and customer better than your customer does, then you would make a move to compete.

Previous to Apple launching the iPhone, BlackBerry exclusively sold through the telecom carrier channel. They didn't in fact, own their customer. The carrier owned the customer. What Apple figured out was that they understood the customer well and that they should and could have a direct relationship with their customer. Apple turned the carriers into a pure infrastructure support system and not a channel per se.

Can you imagine a carrier not supporting the iPhone on their network? The vendor became the customer.

LOOKING AT AI AS BUSINESS MODEL INNOVATION

My vision for the future is an extremely optimistic one. One of my favorite quotes is one from John Adams.

"I must study politics and war, so that my sons may have liberty to study mathematics and philosophy. My sons ought to study mathematics and philosophy, geography, natural history, naval architecture, navigation, commerce and agriculture in order to give their children a right to study painting, poetry, music, architecture, statuary, tapestry, and porcelain."

It puts me in the right mindset of how I think about technology in our lifetime and our children's lifetime. AI, and the effects of its innovation, are already here—this is an unavoidable fact. As we look at the future through the prism of AI, we have a clear choice—in the face of evolution, can you adapt?

In any industry, the most successful and forward-thinking leaders are not the ones who decry the current set of circumstances and wish to return to a previous bygone era. They internalize that change is coming, and so they put plans and processes in place to be successful in this new era. This will require forward-thinking and inventive solutions.

As we've discussed throughout this book, "doing nothing" does not qualify as an inventive solution. If leaders do nothing, they are doing a disservice to themselves and their children, because they will be unprepared for a new AI-driven world. As I have enumerated through these pages, with the right perspective, AI does not have to be a soul-crushing, layoff-producing, technological bogeyman. We simply have to be intentional and lean into the human skills and qualities that algorithms cannot do.

I was recently at an Industry Trade Show event giving a keynote on the future of an AI-driven future.

At the end of the talk during the Q&A session, I had several people stand up and aggressively voice their concerns that AI must be stopped. "As a father of three sons, how can you sleep at night supporting AI? AI is going to take away their future."

I have never been attacked in such a manner, which tells me that the large majority are afraid. In Silicon Valley, the conversations are less about AI being the biggest thing since the Internet. Rather, in technology-advanced markets like Silicon Valley, the focus is that AI can only compare to the invention of electricity. We live in one of the most transformative times. However, leadership is required to help people see beyond fear to see a clear opportunity.

AI is about business model innovation—it is not about technology. These business models will not be invented by AI, they will be invented by people. Not all people, but people with a growth mindset are willing to build the skills to thrive. These skills will not be static; a year from now there will be a whole new set of skills required.

Much like electricity, individual leadership requires that we have the time to "study." AI is a core capability for people to shape the future of the world that they find important and that they are passionate about.

Unfortunately, the early days will focus on automating the work and tasks that we know.

However, this is a shortsighted view and not the correct way to implement AI's possibilities. Yes, AI produces efficiency, but it doesn't generate critical organizational assets like leadership, culture, and ethos. These qualities will be the real drivers for growth in a world where operations are commoditized. Being engaged with markets and society will shape opportunities that we don't even know exist yet. AI will give us the liberty to shape the life that we want versus the life choices that have been given to us.

My hope for this book is that it contextualizes how to think about the changes that will happen as a result of AI. My hope is that it takes away much of the fear and anxiety readers have—in the same way that I've tried to reassure conference attendees about the possibilities of AI rather than just the perceived negative effects. AI is often referred to as a "game changer" in the same way that the Internet was. I tend to disagree—it is as disruptive as electricity. Yes, electricity provided a framework that made numerous other tasks much easier. More importantly, it had a profound positive effect on a number of other industries like manufacturing, appliances, electronics, transportation, entertainment, and communication.

The innovation that electricity spurred in these other fields could be another book entirely. Factories had the ability to reduce costs and produce goods on a much larger scale. New inventions like refrigerators, ovens, and vacuums made domestic tasks far easier. In addition to live performances at the theater, people were also entertained by radio, motion pictures, and eventually television. Communication through the telegraph and telephone meant that people could communicate far more quickly than simply through traditional letters carried by horses and trains.[21]

I don't know about you, but I have no interest in hanging my laundry on a clothesline, handwashing dishes, or reading by candlelight. Yes, as with any technological evolution, electricity made some jobs obsolete—like lamplighters, ice cutters, and coal stokers—but look at how many new careers were born out of it. AI will have a similar effect.

Simply put, if AI is the new electricity, then we are all very fascinated by the light bulb. ChatGPT is the first iteration of AI—we simply have no idea what innovations and new industries it will spur in the coming years. Can you imagine Thomas Edison and his team witnessing the highly illuminated Sphere in Las Vegas in 2024, all thanks to their "electric lamp" patent in 1880?

Ultimately, the future of AI will be written not by the technology itself, but by the humans who implement it.

APPENDIX

1. ChatGPT, response to "Examples of evolution of desktop software programming," Poe by Quora, February 2, 2024

2. ChatGPT, response to "Examples of evolution of visual programming," Poe by Quora. February 2, 2024.

3. ChatGPT, response to "Examples of evolution of AI programming," Poe by Quora, February 2, 2024.

4. ChatGPT, response to "Why doesn't money ball work any more in baseball?" Poe by Quora, February 27, 2024.

5. ChatGPT, response to "What are Milestones in AI since before 2000?," Poe by Quora, February 2, 2024.

6. ChatGPT, response to "What are the types of AI technologies?," Poe by Quora, February 2, 2024.

7. "Robotic Vacuum Cleaner Market Size, Share & Trends Analysis Report," Grand View Research, accessed February 4, 2024, https://www.grandviewresearch.com/industry-analysis/robotic-vacuum-cleaner-market

8. ChatGPT, response to "Give a simple explanation and example of how AI is being used in healthcare diagnostics," Poe by Quora, February 27, 2024.

9. ChatGPT, response to "What is a good example of AI being used in agriculture in a simple tone?" Poe by Quora, February 27, 2024.

10. ChatGPT, response to "How is AI impacting self-driving cars and what is a good example?" Poe by Quora, February 27, 2024.

11. Macht, Joshua. "The Management Thinker We Never Should Have Forgotten." *Harvard Business Review.* June 24, 2016. https://hbr.org/2016/06/the-management-thinker-we-should-never-have-forgotten

12. ChatGPT, response to "Summarize the book the discipline of market leaders," Poe by Quora, February 27, 2024.

13   ChatGPT, response to "Summarize the background of the Matrix and the concept of taking the red pill vs. the blue pill," Poe by Quora, February 27, 2024.

14   "Has iOS killed off Facebook ads? (and why that may be a good thing)" Crimtan: Intelligent Lifestyle Marketing. https://www.crimtan.com/blog/has-ios-14-killed-off-facebook-ads-and-why-that-may-a-good-thing/

15   "The Birth of a Refreshing Idea." The Coca-Cola Company. https://www.coca-colacompany.com/about-us/history/the-birth-of-a-refreshing-idea

16   Dweck, Carol. *Mindset: The New Psychology of Success.* New York: Ballantine Books, 2007.

17   Dweck, Carol. *Mindset: The New Psychology of Success.* New York: Ballantine Books, 2007.

18   Wunker, Steve. "Amazon's Formula for Asymmetric Competition." *Branding Strategy Insider.* https://brandingstrategyinsider.com/amazons-formula-for-asymmetric-competition/

19   Wunker, Steve. "Amazon's Formula for Asymmetric Competition." *Branding Strategy Insider.* https://brandingstrategyinsider.com/amazons-formula-for-asymmetric-competition/

20   Wunker, Steve. "Amazon's Formula for Asymmetric Competition." *Branding Strategy Insider.* https://brandingstrategyinsider.com/amazons-formula-for-asymmetric-competition/

21   ChatGPT, response to "How did the widespread adoption of electricity have a positive impact in numerous industries?" Poe by Quora, February 18, 2024.

ACKNOWLEDGMENTS

You would think that, after publishing two previous books, I would recognize how much work the process of writing a third one involves. Unfortunately, I have the memory of a goldfish.

Besides the family time that gets consumed, it also requires my team to support me. Both my near-field and far-field team.

I would like to thank the entire Shadow team. Specifically, Matt Ohlman who supports all my crazy ideas and talks me out of a fair share of them.

To my industry friends, who continue to "get me," even when I think they won't. I wish I could list you all but, for fear of listing you in the wrong order or omission, I will not.

To all the founders that I work with on a daily basis that endure my advice. The founders are my *why* I get up and show up everyday.

This intertwined journey that we are all on to make the world a better place for future generations is a tough one.

Like my two previous books, I write because I see fear and confusion in the eyes of people within the industry, at conferences, and through other daily interactions. My goal is to help in whatever way I can. I always hope that a word, a sentence, or even a book may help them on their journey.

I truly appreciate y'all. Onwards...

ABOUT THE AUTHOR

K P Reddy is the founder and CEO of Shadow Ventures and and a prominent figure in the world of technology and innovation. With a background in civil engineering from the Georgia Institute of Technology, Reddy has become a global authority in AEC environments, AI, robotics, automation, mobile applications, and cloud computing.

He is the author of *What You Know About Startups is Wrong*, debunking eleven popular myths about what defines a top entrepreneur, and *BIM For Building Owners and Developers*, the definitive guide to building information modeling.

His extensive experience includes founding and exiting three technology companies to NASDAQ, NYSE, and privately held futures, running Enterprise Transformation at Gehry Technologies, and being the General Manager of ATDC at Georgia Tech, one of the oldest technology incubators in the country.

Reddy has appeared on stage at Autodesk University, SXSW, TechCrunch Disrupt, NYC Building Congress, Foreign Direct Investment World Forum, and BuiltWorlds. He's a frequent lecturer at universities including NYU's Stern School of Business, Georgia Tech, Columbia University, and Harvard GSD.

He splits his time between Atlanta and Asheville with his wife and three sons.

ABOUT WHAT YOU KNOW ABOUT STARTUPS IS WRONG

B eing an entrepreneur seems like the ultimate American dream. But the startup mythology is filled with urban legends and false expectations that can cause business owners to lose what is truly important to them. To succeed in the world of entrepreneurship, one should know how to enter with clear goals and strong boundaries.

In *What You Know About Startups Is Wrong*, K.P. Reddy explains the behind-the-scenes reality of business building, debunking eleven popular myths about what defines a top entrepreneur. With guidance on everything from how to balance work and personal life to leveraging great relationships to when and how to leave the game, Reddy guides potential founders and established entrepreneurs alike through the realities of running a company.

Being an entrepreneur isn't a short-term gig with big wins—it's a lifestyle. This book is your essential guide to understanding, preparing for, and managing your business without sacrificing your health, happiness, or relationships.

OTHER TITLES FROM RIPPLES MEDIA

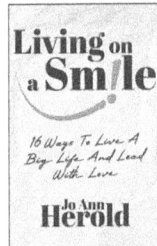

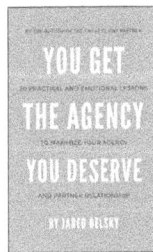

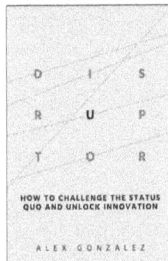

THE 5-DAY TURNAROUND

"It's a tactical, inspiring read!"
Neil Blumenthal, CEO of Warby

THE **48** hours

THE 5-DAY TURNA ROUND

Be the leader you
always wanted to be.

JEFF HILIMIRE

By the Best-Selling Author of
THE 5-DAY TURNAROUND
Part of the Transitional Leadership Series

THE CRISIS TURNA ROUND

Lead through crisis and position
your company for strength.

JEFF HILIMIRE

As the bestselling author of
THE 5-DAY TURNAROUND & CRISIS TURNAROUND
Part of the Transitional Leadership Series

THE GREAT TEAM TURNA ROUND

How to catalyze growth using PVTV™
and The Great Game of Business™

JEFF HILIMIRE

WHAT DOES
YOUR FORTUNE
COOKIE SAY?

SO IMPORTANT IT'S BECOME THE
UNIVERSE IS TRYING TO SHARE WITH YOU

ADAM ALBRECHT

"6Ps of Essential Innovation is a likely useful roadmap for
innovators innovating- highly recommended!"
SCOTT D. ANTHONY,

6Ps
OF
ESSENTIAL
INNOVATION

CREATE THE CULTURE AND CAPABILITIES OF A
RESILIENT INNOVATION ORGANIZATION

MICHAEL MCCATHREN

Living on
a Sm le

16 Ways To Live A
Big Life And Lead
With Love

Jo Ann
Herold

By the Best-Selling authors of
THE 5-DAY TURNAROUND & WHAT DOES YOUR FORTUNE COOKIE SAY?
Part of the Transitional Leadership Series

THE CULTURE TURNA ROUND

JEFF HILIMIRE
ADAM ALBRECHT

THE CONQUERING CREATIVE

A Battle to Build at the Unstoppable Creative Business

WILLIAM C. WARREN

YOU GET

20 PRACTICAL AND EMOTIONAL LESSONS

THE AGENCY

TO MAXIMIZE YOUR AGENCY

YOU DESERVE

AND PARTNER RELATIONSHIP

BY JARED BELSKY

D I S
R U P
T O R

HOW TO CHALLENGE THE STATUS
QUO AND UNLOCK INNOVATION

ALEX GONZALEZ